IEE HISTORY OF TECHNOLOGY SERIES 13

Series Editor: Dr B. Bowers

WIRELESS
the crucial decade

History of the
British wireless industry
1924-34

Other volumes in this series:

WIRELESS
the crucial decade

History of the British wireless industry 1924-34

Gordon Bussey

Peter Peregrinus Ltd. on behalf of the Institution of Electrical Engineers

Published by: Peter Peregrinus Ltd., London, United Kingdom

Peter Peregrinus Ltd.,
Michael Faraday House,
Six Hills Way, Stevenage,
Herts. SG1 2AY, United Kingdom

British Library Cataloguing in Publication Data
Bussey, Gordon
    Wireless—the crucial decade: history of the British
    wireless industry 1924-34.
    1. Radio equipment: Receivers, history
    I. Title II. Series
    621.3841809042

ISBN 0 86341 188 6

Printed in Singapore by Stamford Press

To Katherine
and the precious hours together

Contents

Preface

This is a history of the British Wireless Industry in its formative years. After setting the scene from the late 1880s, it is convenient to start with the technical background, then look separately at broadcasting trends in Britain and continental Europe, British radio valves, receiver development in America and a brief look at Germany and France, before embarking on the main theme of British domestic wireless in six sections. All the foregoing chapters are found to have an influence smaller or greater on the History of the British Wireless Industry, 1924-34. The last chapter is devoted to the growth and decline of home construction, with a special reference to kit sets with which, as 1939 approaches, the story ends.

Acknowledgments

I would like to express my thanks to Mr Guy Hartcup for reading the manuscript of this book, and for his helpful suggestions. The original typescript was also read by Dr Brian Bowers, Mr Douglas Byrne, Mr Donovan Dawe, Mr Pat Leggatt and the Rev Colin MacGregor. I am grateful to them for their encouraging comments. Among others who have helped or advised in this production I would like to thank: Mr Keith Geddes, Dr Tom Going, Mr Robert Hawes, Mrs Pat Murphy, Miss Pat Spencer, Mr Richard Williams and the management of Philips Electronics.

I am indebted to Electronics World, successor to Wireless World, an invaluable source of both text and illustrations.

I want to thank Mr Ian Robertson for his generous help. He freely gave time, information and encouragement, read the manuscript and made valuable suggestions.

Gordon Bussey
Purley
January 1990

Chapter 1
The scene to 1924

In 1888, just over one hundred years ago, Heinrich Hertz first carried out controlled experiments to transmit and detect high frequency electromagnetic oscillations through space, thus verifying experimentally James Clerk Maxwell's theoretical predictions made some twenty years previously. Further studies using spark-generated waves of some 30cm length allowed him to confirm the chief predictions of the theory and also showed the importance of bringing the detector into resonance with the transmitter, 'tuning' in fact. These primitive experiments were the first steps on the road leading to wireless telegraphy and hence broadcasting, a road along which the amateur experimenter has played a continuous part. Following Marconi's pioneering work at the turn of the century, and aided by the discovery of the Edison Effect leading to the Fleming Diode, modified by de Forest, continuous waves of relatively constant frequency could be generated, transmitted, intercepted and detected by 1913.

A powerful stimulus to the urgent development of this new art was undoubtedly the 1914–18 war, when production of valves in particular and other wireless apparatus in general was engineered on a grand scale; large numbers of people, and especially service personnel, were introduced to the new technology. The Presidential Address, by Dr. W. H. Eccles, F.R.S., to the Radio Society of Great Britain (R.S.G.B.) on 26 September 1923 refers to this '. . . when the war came, the amateurs penetrated in their hosts into the armies, and turned their wireless experience and their talents to the design, construction, operation and improvement of apparatus for use in war'.[1] Wireless World itself commented on this, too, '. . . the valve had arrived, and the stay-at-homes envied the service men their knowledge of it'.[2]

Following the chaos of war, events moved at an accelerating pace and by 1922, besides the regular commercial transmissions for shipping, meteorology etc., some informal 'broadcasting' was available to delight the energetic band of experimenters. As a result of considerable pressure by companies interested in developing a wireless industry, ably supported by an influential band of experimenters, a 'Broadcasting Petition' signed

204   THE WIRELESS WORLD AND RADIO REVIEW   NOVEMBER 11, 1922

The New Licence for Broadcast Reception

BROADCAST   LICENCE.

A 41602

WIRELESS TELEGRAPHY ACT, 1904.
Licence to establish a wireless receiving station.

Mr. *The Wirelessworld & Radio Review*
(Name in full)
of *12/13 Henrietta St. W.C* ....is hereby
(Address in full)
authorised (subject in all respects to the conditions set forth hereunder) to establish
a wireless station for the purpose of receiving messages at *Henrietta Street*
(address)
....for a period ending on the ....next.
(Date of expiration.)

APPARATUS USED UNDER THIS LICENCE MUST BE MARKED

The payment of the fee of ten shillings is hereby acknowledged.

Dated 3rd day of *November* 1922

Issued on behalf of the Postmaster-General
O.C.N.

Signature of Licensee..............................
WIRELESS WORLD & RADIO REVIEW.

If it is desired to continue to maintain the station after the date at which this fresh
Licence must be taken out within fourteen days. Heavy penalties are prescribed by the
Wireless Telegraphy Act 1904, on conviction of the offence of establishing a wireless
station without the Postmaster-General's Licence.
2801 G & S **194**

Stamp of Issuing Office.

CONDITIONS.

    1. The Licensee shall not allow the Station to be used for any purpose
other than that of receiving messages.

    2. Any receiving set, or any of the following parts, vizt.:—Amplifiers
(valve or other), telephone head receivers, loud speakers and valves, used under
this licence must bear the mark shewn in the margin.

    3. The Station shall not be used in such a manner as to cause interference with the working of
other Stations. In particular valves must not be so connected as to be capable of causing the aerial
to oscillate.

    4. The combined height and length of the external aerial (where one is employed) shall not
exceed 100 feet.

    5. The Licensee shall not divulge or allow to be divulged to any person (other than a duly
authorised officer of His Majesty's Government or a competent legal tribunal) or make any use
whatsoever, of any message received by means of the Station other than time signals, musical
performances and messages transmitted for general reception.

    6. The Station shall be open to inspection at all reasonable times by duly authorised officers of
the Post Office

    This Licence may be cancelled by the Postmaster-General at any time either by specific
notice in writing sent by post to the Licensee at the address shewn hereon, or by means of a general
notice in the London Gazette addressed to all holders of wireless receiving Licences for broadcast
messages.

    N.B —Licences may only be held by persons who are of full age, and any change of address
must be promptly communicated to the issuing Postmaster

2801

At the top will be seen a reproduction of the front of the licence.   The Conditions are on the back of the form.

by sixty-four societies representing most amateurs was given to the Postmaster General on 29 December 1921. Following this, awakening interest of the population at large was catered for by public service broadcasting becoming an accomplished fact on 14 November 1922. The station was 2LO, the London station of the British Broadcasting Company, and the first broadcast was the news read by Arthur H. Burrows. The Company, with J. C. W. Reith (later Lord Reith) as General Manager, expanded rapidly. Birmingham 5IT and Manchester 2ZY opened one day later than London, to be followed by Newcastle 2NO, Cardiff 5WA, Glasgow 5SC, Aberdeen 2BD and Bournemouth 6BM. These eight stations formed the original main stations of the Company and each had a power of about 1.5kW. For reception, a 'Broadcast Licence' had to be obtained and only receiving apparatus bearing the 'B.B.C Type Approved by Postmaster General' stamp could be legally used, as revenue for the funding of the Broadcasting Company came from this source.

Some indication of the interest in wireless matters shown by both manufacturers and public can be judged from the success of the First All-British Wireless Exhibition held in the Horticultural Hall, Westminster, during the first week of October 1922. Some sixty stands were taken by companies anxious to display their wares and a record number of paying visitors attended, over 6000 on the Wednesday alone with an average of 5000 a day for the whole period. At the opening, Sir Henry Norman M.P., who had played an important part in introducing wireless to the armed forces both before and during the war, made an unusually prophetic speech: he believed 'that broadcasting was destined to become an integral part of our everyday life as the ordinary telephone is to-day. Broadcasting, . . . was no passing craze. . . . in a year or so it would have become so commonplace a necessity that it would cease to be mentioned'.[3] That the interest was nationwide is shown by a glance through some half-dozen issues of Wireless World in the autumn of 1922 where over a hundred wireless societies were reported ranging from Aberdeen to Highbridge and Burnham-on-Sea and from Rhyl to Scarborough.

With the introduction of broadcasting, came the joys of 'listening-in' – 'And when you tuned in you had no mind to be choosy – even if it had been possible – about what programme you heard. To hear even a noise was an achievement. The scratching of your "cat's whisker" on the crystal was music itself'.[4] Most listeners used crystal sets, indeed B.B.C. policy was to place their transmitters so that the majority of the population was within 'crystal range'. To this end the Chief Engineer, P. P. Eckersley, filled in the biggest gaps by a system of 'relay stations' of low power; they were 'relays' in the sense that they had none of the studio facilities of the main stations but simply re-radiated a programme from one of them. An interesting quotation occurs in a much later edition of Wireless World 'Radio Show in Retrospect'. Describing conditions in 1922 it runs – 'Bearing in mind the fearsome appearance of valve receivers at this time, it is not to be wondered that the simple crystal set received the greatest support. . . . The only outstanding receiver designed exclusively for broadcasting was the Marconiphone V2'.[5] This set was advertised at a price of £22.8.0 complete, and

OCTOBER 10, 1923   THE WIRELESS WORLD AND RADIO REVIEW   xxiii

POST OFFICE NOTICE.

USE OF UNLICENSED WIRELESS RECEIVING SETS.

The Postmaster-General calls attention to the new arrangements announced in the Press for the issue of wireless licences.

Many persons are known to be using wireless receiving sets without a licence, owing to the fact that no licence has hitherto been available for home-made sets. A new form of licence known as an "Interim Licence" has now been introduced to meet the case of persons **who are already in possession of unlicensed sets.** It imposes no condition as to the make of existing apparatus.

This licence will be issued at an annual fee of 15s. to persons who apply **before the 15th October.** No charge will be made for past user, and no proceedings will be taken in respect of past user if the licence is applied for before the 15th October. **Any person who uses unlicensed apparatus after that date will render himself liable to heavy penalties under the Wireless Telegraphy Act, 1904.**

The "B.B.C." Licence at 10s. still remains on sale, and a second new form of licence, known as a "Constructor's Licence," which will meet the case of persons who intend to make their own sets but have not yet done so, is also issued at 15s.

The new licences are on sale at all Head and Branch Post Offices and certain Sub Offices. Forms of application can be obtained at any of these offices and also at any Sub Office at which Money Orders are issued.

a map of Europe was displayed showing places from which 2LO could be heard on a V2, the furthest distance being St Vincent (Portugal), 1125 miles away.[6]

Many other valve and crystal sets were available and some idea of their range may be gleaned from an advertisement. This lists various McMichael receiver equipment and quotes a complimentary letter from a Captain Grant of Glenlivet, Banffshire,[7] who had purchased items as follows:

Receiver: 4 valve detector amplifier MH4R   £25. 0. 0.
           Tuner Panel   MH/LC   £ 8. 0. 0.

Captain Grant would also have required, typically, these accessories:

Coils		£ 2. 7. 3.
Battery	HT	15. 0.
	LT	£ 1.10. 0.
Aerial 100ft		6. 6.
Insulators 4		4. 0.
Headphones 2000 ohm		£ 2.15. 0.
Valves 4 @ 17/6 each		£ 3.10. 0.
	TOTAL	£44. 7. 9.

This was at a time when an 'experienced teacher to take charge of wireless college'[8] had a salary of £200 a year plus commission and a 'Strong Lad' of 16 seeking a situation 'connected with wireless or electrical trades, any capacity'[9] was willing to work for 15/- a week. Perhaps more in keeping with incomes like these would be a Fellows 'Fellocryst' crystal set complete with headphones and aerial equipment at £3.7.6 + 1/6 for postage.[10]

With prices like these it is easy to see why home construction by amateurs and enthusiasts might be favoured. Building your own wireless was also fostered by the widespread availability of magazines, like Wireless World, which, for more technically advanced readers, ran an extensive 'Questions and Answers' section where specific wireless problems were discussed and answered. Wireless World was also sensitive to the needs of absolute beginners. Full constructional details were given for a complete crystal receiver.[11] The components, including all accessories, could be purchased from advertisers in the same edition of the magazine for £1.12.6, thus giving equivalent results for less than half the price of the Fellocryst. The differences were more remarkable in valve sets; to construct a four-valve equipment comparable with Captain Grant's McMichael (using ex-Government equipment where possible) was less than £10. Particularly interesting in this context is an advertisement for the 'Wireless College, Colwyn Bay'[12] offering a complete kit of parts for a four-valve set, excluding valves, for £18.0.0. It is noteworthy that the position of the amateur builders was recognized officially, special 'experimenters' licences could be bought, and in fact by the end of February 1923 more than a third of all licences were for experimenters. 'Up to 28 February . . . 56,000

broadcast and 30,000 experimental receiving licences had been issued'.[13]

And yet Susan Briggs in 'Those Radio Times' could write: 'The high point of the wireless constructionist had already been reached by 1926'. She continues, suggesting a reason: 'As the valve set playing through loudspeakers replaced the simple crystal set and earphones, the necessary new technique went beyond the skill of most amateur "hobbyists". The novelty of wireless had, in any case, worn off. As Reith had hoped, interest in the programmes broadcast began to replace interest in the "miracle" of broadcasting'.[14]

Two quotations from Reith himself graphically illustrate the domestic wireless scene of 1924. The first is from an article written for 'Popular Wireless'. Speaking about wireless experimenters, he says, '. . . Wireless . . . is the hobby, perhaps par excellence, which is useful and profitable alike, to him who indulges in it and to those who live with him . . . the instrument in its crudest form is complete and a source of satisfaction and of interest and pleasure'.[15] The other is from an article in 'The Radio Times', of May 1924, quoted by Susan Briggs, '. . . we lead our friends into a room where there obtrude on the attention wires and valves and boxes and switches, and, to crown all, a horn. The attention is distracted by all this paraphernalia and by the tuning preliminaries which ensue. And then we all sit with our eyes glued to the loudspeaker and come to the conclusion that the sound is metallic and unsatisfying and that we do not like our music tinned'.[16] The picture presented is not in any way different from that of 1922. Changes had, however, taken place in broadcasting, some of which foreshadowed later events.

In 1922 the B.B.C. had only three of its transmitters at work. By the start of 1925 it had nineteen on the medium waveband, nine main stations of 1.5kW, and ten relays of 120 watts. The main stations had been expected to provide a reliable service radius of about seventy miles but in practice, with existing interference levels, the safe limit was about thirty miles for valve sets, ten miles for a crystal. Relays had a range of about ten miles for valve sets, perhaps three for a crystal. These transmitters served the majority of populous districts but many larger areas could hardly be said to be served at all. To try to fill in these gaps the B.B.C. decided to try a powerful central transmitter operating on long waves and thus less susceptible to long distance night-time fading. The experimental station 5XX was first sited at the Marconi works, Chelmsford, where it was built and moved to Daventry in July 1925. From its inception this station was a success and ensured national coverage. On the Continent also, similar advances in transmitter power had taken place. Wireless World described the overall situation;[17] about seventy European stations were operational, only one, (in Portugal) having a power of much over 1 kW. Apart from station 5XX, both British and European broadcasting had spread by a steady increase in the numbers of similar low power transmitters. The success of this policy in Britain could be measured by the rapid growth in licences; 'Amateur Wireless' carried a note to the effect 'Of the million-odd licence holders in Great Britain today, approximately 65% are crystal users'[18] – taking low power stations to the populous centres, allowing

The New High-Power Station

The Chelmsford works of the Marconi Company where 5XX, the Experimental High-Power Station of the British Broadcasting Company, is situated.   The huge masts of the steel tube type are four hundred and fifty feet high.

relatively cheap crystal sets to bring broadcast reception to many people, had been a remarkably successful policy.

No very fundamental changes had taken place in receiving valves either between 1922 and 1924. It is true that more economical 'dull emitter' filaments were more readily available, true that many new valve 'manu-facturers' had come (and some had gone), true that great claims were made for marvellous new types, but the reality was there were many manufacturers producing very large numbers of basic triode valves which showed very little advance on those generally available in 1922. Wireless World lists ninety-six different valves by eleven different manufacturers,[19] but it is plain that many were completely interchangeable – if a different filament battery were used. The only difference over the two years was that prices had fallen. Components also showed only slight improvements over the two years; prices were somewhat reduced and more stable character-istics had been obtained, but again it was 'more of the same' from a bewildering display of manufacturers, large and small. A perusal of the pages of Wireless World for 1924 reveals over one hundred and twenty firms advertising, all vying for business.

The 'Wireless Exhibition', this time held at the Albert Hall early in October 1924 was reported in various journals. Wireless World gave a review and, in relation to components, it stated '. . . the trend of present-day design, . . . would appear to be towards simplification without in any way sacrificing efficiency'. With reference to complete sets it contained the telling phrase 'The usual receiving sets will be prominently displayed',[20] and again 'Radio Shows in Retrospect' for 1924 'Receiving sets in general had undergone a process of structural simplification without any out-standing change in circuit principle'.[21] Thus, apart from some price reductions, there was almost a static situation so far as the set-buying public was concerned.

The needs of the home constructor were well served. The number of journals devoted to wireless matters continued to expand; the number of wireless societies reporting to Wireless World continued to grow. The price of components had fallen, sharply in some cases. Perhaps the reduction was due partly to increased competition, but increased sales must have played a part. The journals carried numerous constructional articles, some with free blueprints and detailed lists of components and suppliers. A simple crystal set could be made up from components for £1, while a four-valve set would be around £15. There was also a growing choice of kits of parts, supplied complete, to make up either published designs or sets peculiar to the supplier. Just one copy of 'Modern Wireless' (October 1924) carried advertisements for: Peto-Scott, Raymond, Burne-Jones and Peter Curtis. Many others could be found in other issues and in other journals. The position of the home constructors, then, had improved from 1922 to 1924, and the financial benefits were, if anything, even greater.

Amateurs, for their part, were beginning to show signs of a polarization of interests. A speech by Dr. W. H. Eccles, F.R.S., to the Institution of Electrical Engineers had as its theme the position of the scientific amateur and contained the passage: 'The Radio Society and its Affiliated Societies

comprise every kind of amateur – the home constructor who has fallen a victim to the fascination of making or improving tuners and amplifiers, the ripe enthusiast who welcomes morse as much as music and constructs reflex circuits and other modern marvels, and the matured transmitter who fishes in the oceans of space. . . The advent of broadcasting has greatly augmented the ranks of the amateurs, especially those interested in building and experimenting with receiving apparatus. . . . It is inevitable that a large proportion of these amateurs will in due time desire to supplement their knowledge of receiving apparatus by actual work with transmitting apparatus . . . and thereby enter the . . . élite in the amateur movement'.[22] As this speech so accurately foretold, so long as the ranks of listeners continued to grow, so would the numbers of home constructors, but some would transfer from their ranks to other branches of the hobby. Home constructors would be expected to reach a peak in numbers at some point. In general, the picture in 1924 is of a gently expanding broadcasting service, leading to a steady increase in the numbers of listeners, numbers of home constructors and, indeed, to a steady increase in activity throughout the wireless world.

Two events in 1924, however, paved the way for a complete transformation; the opening of the high powered station 5XX was to set a precedent which European broadcasting was to follow with chaotic results; the other was the change in the licensing system which came into force on 1 January 1925. From that date there was only one type of licence (costing 10/-), but most importantly the restriction to 'all British' sets, components, valves etc., inherent in the B.B.C. licence, was abolished. No longer did wireless apparatus have to be of British manufacture and so the door to foreign, especially American, competition was opened.

Since America had started public broadcasting two years earlier than Britain, it had advanced well along the road. In addition, the much larger market and general manufacturing capacity had enabled prices to be brought down much more than here. An article by N. P. Vincer-Minter comparing British and American receivers is of considerable interest, several quotations are of relevance '. . . it must be admitted that from the point of view of selectivity, sensitivity, ease of control and variety – not to mention artistic appearance – the palm must be awarded to the Americans . . . the Americans are far ahead . . . in producing sets in great variety of really selective and sensitive "DX" (long range) capabilities'. The writer also refers to the existence of many American sets specifically designed for 'low wavelengths from fifty up to about two hundred metres'. With reference to home constructors in America the article points out: 'It is possible . . . to go into any reputable radio store . . . and . . . be handed an attractive cardboard box containing the complete set of parts according to the author's specifications, . . . arranged in boxes in a similar manner to sets of Meccano parts'.[23] As we shall see, it took another three years or so for this to be possible on any scale in Britain. Another article discusses American reception conditions and indicates the need there for long range and high selectivity requiring in particular the widespread use of superheterodyne receivers.[24]

In spite of America's obvious technological lead and the removal of the barrier to powerful competition from there – and from elsewhere – the editor of Wireless World is quite sanguine of the outcome of foreign competition and noting that: 'The net result, therefore, is likely to be keener competition in the production of really high-grade apparatus at reasonable prices, and. . .the British manufacturer has good reason to feel fairly secure'.[25] Just how far this prediction became true, and just what effect American and other foreign influences were to have on the British scene may appear as the rest of this story unfolds.

What then had happened so quickly to so change the conditions of experimenters? A study of the decade 1924–1934 may throw some light on the complex interplay of technical advances, political pressures and domestic requirements which led to the eventual eclipse of the experimenter. Certainly conditions did change. 'Possibly no boom like that of the Winter of 1923–24 will ever be experienced again by British radio dealers and manufacturers'.[26]

References

1 Wireless World, 10 October 1923, p. 51
2 Wireless World, 25 November 1922, p. 260
3 Wireless World, 7 October 1922, p. 8
4 B.B.C. Scotland '50', Anniversary publication
5 Wireless World, 24 September 1930, p. 302
6 Wireless World, 10 October 1923, p. xxiii
7 Wireless World, 6 January 1923, p. xxxix
8 Wireless World, 27 January 1923, p. liii
9 Wireless World, 31 March 1923, p. xlii
10 Wireless World, 7 October 1922, inside back cover
11 Wireless World, 24 February 1923, p. 685
12 Wireless World, 14 October 1922, p. xxv
13 Wireless World, 31 March 1923, p. 872
14 S. Briggs, Those Radio Times, p. 28
15 Popular Wireless 15 March 1924, p. 79
16 S. Briggs, Those Radio Times, p. 28
17 Wireless World, 27 May 1925, p. 500
18 Amateur Wireless, 14 February 1925, p. 320
19 Wireless World, 30 September 1925, p. 426
20 Wireless World, 1 October 1924, p. 25
21 Wireless World, 24 September 1930, p. 303
22 Wireless World, 1 October 1924, p. 6
23 Wireless World, 4 February 1925, pp. 601ff
24 Wireless World, 27 May 1925, pp. 498ff
25 Wireless World, 31 December 1924, p. 431
26 The Wireless and Electrical Trader, 11 March 1944, p. 288

Chapter 2
Technical background

'"The trouble with wireless", a clergyman once told me, "is that there is so much theory about it"', John Scott-Taggart.[1]

At the transmitter, the radio signals with which we are concerned have two main elements (1) A High Frequency (H.F. or R.F.) radio carrier wave generated by a valve oscillator (few of the old 'spark' transmitters survived to intrude for long into the period under review). (2) The Low Frequency (L.F.) audio information impressed on this carrier. The amplitude (strength) of the radio frequency carrier was varied in response to the required audio frequency wave, a process termed 'amplitude modulation' (A.M.).

At the receiver, there were again two elements (1) A tuned aerial system to intercept the wanted radiated signal; a coil of inductance L and a capacitor of capacity C in circuit together resonate (tune) to only one frequency f, given by $f = \frac{1}{2\pi\sqrt{LC}}$ varying L and/or C changes the resonant frequency f, thus the one required signal can be selected out of the multitude present. (2) A detector to extract the low frequency (audio) modulation. A suitable crystal (or crystals) or a valve (in one of a specific number of circuits) 'detected' the incoming R.F. signal by rectifying it, i.e. passing only the positive, or the negative, half cycles of the radio wave and allowing the wanted audio frequency modulation to be reconstituted (rendered audible in headphones or loudspeaker).

The ability of a tuned circuit to be selective (to separate neighbouring transmitting wavelengths) and to be sensitive (passing on a strong signal to the detector) are both expressed in terms of the 'Q' of the circuit (Quality Factor).

For a coil of inductance L and operating at a frequency f, Q is expressed by $Q = \frac{\omega L}{R}$ where R is the 'radio frequency resistance' of the coil. This serves as a definition for the efficiency of a complete L-C tuned circuit; it is really a measure of the energy losses on the circuit. If Q be high, (say 100) then the circuit is 'sharp tuned', i.e. it responds only to a narrow band of frequencies close to resonance, and the signal passed on to the detector is strong. A low Q (10 or under) has the opposite effect. High Q can usually be ensured by a large diameter coil, kept well away from any other conducting surface (metal), wound with reasonably thick wire of high conductivity and with a minimum of insulating material present as this is often a source of losses.

The normal 'broadcast bands' used were:

(1) Medium wave (called 'short' in early days), from:
$\left\{\begin{array}{l}\text{300m (later 200m) up to 600m}\\ \text{1000kc/s (later 1500kc/s up to 500kc/s)}\end{array}\right\}$ M.W.

(2) Long wave, from:
$\left\{\begin{array}{l}\text{1000m to 2000m.}\\ \text{300kc/s to 150 kc/s}\end{array}\right\}$ L.W.

here we use the conversion given by:

Wavelength (in metres) $\times$ frequency (in c/s) = c

Where c is the speed of light = 3×10^8 m/s

The normal audible band (L.F.) runs from near 50 c/s to 15,000 c/s or above, depending on the individual, age, etc.* Modulating an R.F. with an audio frequency has the effect of 'spreading' the R.F. carrier into a range of frequencies (sidebands) centred on the original frequency. The 'width' of the band of frequencies is twice the highest modulating frequency. In the period under review, the audio frequency modulation was restricted to frequencies up to about 4500 c/s as this enabled more transmitters to be accommodated within the confines of the broadcast bands. A transmitter of, say, 200 kc/s (1500 m, LW) would thus occupy a band of frequencies from 200 kc/s − 4.5 kc/s to 200 kc/s + 4.5 kc/s, i.e:

{195.5 to 204.5 kc/s; expressing the frequency
{1534 m to 1465 m; expressing the wavelength.

To receive this transmission, a wireless set required a tuner circuit capable of responding equally well to all frequencies within this band and to none outside it. This is not achievable in practice but can be approached.

Having discussed the requirement of the transmitted signal, we can now discuss how various receivers may be arranged to make use of a transmission.

Whatever the style of the receiver, a tuned input circuit is essential. While older sets tune the aerial, later circuitry loose coupled the aerial to the input stage tuned circuit. This of course was a considerable advantage since any aerial could be used without affecting receiver tuning. In an L-C tuned circuit, either or both L and C could be varied to allow the circuit to resonate at the wanted frequency.

(1) *Variable inductance, L*
   (a) *Variometer*: two coils arranged to move relative to each other, the interaction between the two increasing or decreasing L.
   (b) *Tapped coil*: coil of many turns 'tapped' by fixed contacts at points along the winding; a switch or switches varied L by tapping off varied numbers of turns.
   (c) *Slider tuning*: long coil of many turns; sliding contact allows larger or smaller numbers of turns to be selected.

*The modern unit for frequency is Hertz (Hz) rather than the earlier cycles per second (c/s).

(*d*) *Spade tuning*: flat coil with metal spade (or disc) arranged to be moved across it, L is decreased the more metal is present.

For all these circuits, the required tuning capacitance was often that present incidentally as 'stray' capacitance, aerial capacitance etc.; this was thought to be advantageous since the Q of a tuned circuit was greater for a greater L. Usually for a long external aerial a small series capacitor was introduced between aerial and tuning circuit.

(2) *Variable capacitor, C*
An adjustable capacitor is used in series or in parallel with an inductance.

The useful tuning range of a particular coil was limited. In early days interchangeable plug-in coils were used. Later, switching became possible without excessive losses. In every case the single tuned circuit had a low Q so that there was no difficulty in passing all the sidebands. In fact, the tuning was so broad that it was difficult to separate stations on different frequencies. Spade-tuning in particular was very inefficient and led to very broadly tuned circuits.

A crystal detector drew its energy from the tuned circuit and therefore introduced heavy 'damping', resulting in very broad-tuned, inefficient, resonant circuits. In consequence, the useful range of a crystal set was small and it was not in a position to separate two stations if their operating frequencies were at all close.

A valve detector fared rather better, detector damping was less and so the tuned circuit was sharper; the valve amplified the resultant L.F. signal perhaps 5 or 6 times, so the range of a simple valve set was substantially greater. The major advantage of the valve receiver lay in 'reaction' however. Here a small amount of the R.F. signal (which was also amplified by the valve) was fed back from the output (anode) to add on to the input (grid). The valve could thus draw power from the H.T. battery and feed it back to the tuned circuit to cancel out the energy losses there, thus making the Q much higher, with much sharper tuning and a more powerful signal. Single valve sets with reaction like this could, under favourable conditions, give reception over distances many times the maximum achievable by a crystal set. There was one difficulty; over-use of reaction caused the valve to oscillate – acting like a low power transmitter. This disturbed reception by other sets over a large area. Such reaction back to the aerial circuit was at first forbidden under the B.B.C. rules, though this ruling was subsequently relaxed. It is noteworthy that, although reaction could be used to sharpen up tuning because of the greater Q, this inevitably cropped the wanted side-bands and quality suffered.

Further L.F. amplification could be obtained by feeding the output from the detector, whether it be crystal or valve, to the grid of another valve – and yet again – to give two or more L.F. stages. The coupling in early days was almost invariably performed by transformer, as this could be arranged with more turns in the secondary than the primary, so the voltage passed on from one valve to the next was amplified: amplification of 3, 5 or more times was common in primitive sets. Since the 'gain' of a valve was unlikely to be more than 5, use of 'step up' transformers markedly increased the loudness of the sounds. Unless such transformers were very well designed, they introduced a great deal of distortion.

Attempts were made to amplify the radio frequency signal from the tuned circuit before passing it on to the detector. The triode valve was not satisfactory for this purpose, however. The anode and grid are two metal conductors separated by an insulator (the evacuated space in the valve) and so act as a capacitor. This always fed back some of the amplified signal from the anode to the grid, a process known as the 'Miller Effect'. The effect was much worse when the valve was operating at high frequencies with the result that triode valves could usefully amplify only very low R.F.; 30 kc/s was possible, with care, but 100 kc/s was very difficult. All sorts of expedients were tried, the commonest being to use very inefficient coils, or to cause the valve to draw energy from the tuned circuit and thus damp it. With care a stage gain of perhaps 1 or 2 could be obtained at about 500 kc/s. The real advantage of such triode R.F. stages lay in the ability to use tuned circuits in the couplings between valves, and thus enhance the sharpness of tuning without introducing unacceptable signal losses. By judicious manipulation it was possible to produce a circuit which would give a useful gain, would sharpen up the tuning, and would remain adequately stable. Usually, any attempt to shift the tuning to another station resulted in instability and oscillation.

One circuit was devised which did allow for stable R.F. amplification by triode valves. This was the so-called neutralized-triode. Here a small capacitor was connected between the grid and the output deliberately, but it was so arranged that it fed back a signal which just cancelled out that due to the inherent grid/anode capacitance. A neutralized amplifier was difficult to keep in alignment over the whole tuning range, but it could be made to give a gain of 10 or 12 in signal strength. Very often gains of more than this were achieved, but they were always due to uncontrolled reaction occurring.

In 1927–28 stable H.F. amplification at last became possible, as the screen-grid (S.G.) valve was introduced. This had a screen between anode and grid which, with suitable circuitry, reduced the grid/anode capacitance to negligible proportions. Using later versions of these valves, stage gains at R.F. of several hundreds were possible with complete stability.

During the late 1920s a 3 or 4 valve set became almost standard, consisting of an H.F. amplifier, a detector and one or two L.F. amplifying stages (H.F. det L.F.); by using an S.G. valve with tuned aerial and tuned coupling to the detector – and with careful use of the ubiquitous reaction control – stable amplification could be obtained along with adequate selectivity. The detector was usually coupled with a 3:1 or 5:1 transformer to a power valve which was able to drive a loudspeaker. Such a set was known as a T.R.F. (Tuned Radio Frequency) set. It was sometimes called a 'straight' set as the required signal was fed stage by stage through the set without altering the frequencies contained in the received transmission. After 1927–28 the Pentode valve became available and allowed for a much greater L.F. gain; in some sets advantage was taken of this to dispense with one of the L.F. stages (quite often there were two) or, occasionally, to abandon the coupling transformer and use Resistance/Capacity coupling in its place, so improving quality and reducing cost.

Throughout the 1920s and 1930s, more, and more powerful, trans-mitters came into use and selectivity of a much higher order was required. This could be done by putting in two R.F. valves, with three tuned circuits, but each additional tuned circuit required an additional tuning knob, making the receiver again more like a piece of laboratory apparatus than something suitable for a layman to use. One particular set was a Marconiphone 81 – also called the 'straight 8' – which had no fewer than six quite independent tuning knobs, each of which had to be set to the exact frequency before reception was possible.

American sets of the period 1925–26 usually had 3 or 4 H.F. valves, but they arranged for all, or often all but the aerial circuits, to be tuned by a 'ganged' capacitor – several separate tuning capacitors mounted on one axle and all rotating together. In order to make this system workable, the tuned circuits had to be made broad-tuned, otherwise the inevitable circuit differences – stray capacities, slight differences in the coils, valves, capacitors – made the correctly-tuned circuits deviate from each other as the tuning control was moved, causing marked variations in efficiency at different points on the dial. This effect could be masked by making the circuits broad-tuned, i.e. low Q, and thus throwing away much of the value of extra tuned circuits. In a practical sense these unwanted variations made it very difficult to preserve anything like a 'pass-band' of 9 kc/s at all points over the required tuning range. In early sets this did not matter much as loudspeakers were of such poor quality and produced little more than 'a jolly, cheerful noise'. The old horn loudspeaker with a taut iron diaphragm was full of resonances. The later 'moving iron' type, attached to a paper cone, was better, but the vibrating iron driving-mechanism also had very marked resonant frequencies which introduced distortions. The moving coil loudspeaker gradually came into use in the late 1920s and had fewer defects; it was sufficiently better than its predecessors to make it imperative to improve the circuitry and the tuning arrangements to make full use of its capabilities. In a modified form it is still in current use.

A different solution to the twin problems of selectivity and bandwidth was provided by the supersonic heterodyne or superhet for short. The basic idea is simple and had a very long pedigree going back to the early days of wireless, but it led to two distinct generations of superhets reflecting changes in available technology.

For both, the theoretical principle is the same. The incoming required modulated Radio Frequency is intercepted by an aerial system, as usual tuned to receive it. This original frequency is passed on to a valve where it is 'mixed' with a locally-generated, unmodulated, oscillation. In the mixer valve, often called the frequency-changer, 'sum and difference' frequencies are generated. The process is best illustrated by an example. Suppose we wish to receive an incoming frequency of 200 kc/s. If the local oscillator is set to 150 kc/s and the two frequencies are mixed in the one valve, then in the anode circuit will appear the two original frequencies together with new frequencies equal to the sum and difference of these two. The anode circuit is tuned to select only the different frequency (200 − 150 = 50 kc/s) so that this 50 kc/s signal alone is passed on for further amplification. All subsequent amplification up to the detector would be carried out at this fixed 'Intermediate Frequency' (I.F.). To receive a station having a different frequency, say 240 kc/s, the aerial circuit is tuned to this new frequency, the local oscillator is also retuned, to 190 kc/s, then (240 − 190) − 50 kc/s again.

The main amplification is thus carried out at this I.F., and since it is a fixed frequency, any convenient number of amplifying stages can be used without any problems of 'ganging' associated with variable frequency amplifiers. There is another inherent consequence. If tuned circuits of high Q are used the selectivity is very high; however, the sidebands are cut and quality is poor, but if a number of sharply tuned circuits are 'stagger-tuned', i.e. set to slightly different frequencies, it is possible to construct an amplifier such that it corresponds more or less closely to the ideal of passing a 9 kc/s-wide band centred on the I.F. and with a very sharp cut-off on either side. This could be obtained in an I.F. amplifier, as the careful stagger-tuning could be done once and for all at the factory and remain in alignment for many years. In the older generation of superhets, the band-pass was not of much importance as quality was always poor, but the ability to transform any wanted signal to a fixed low intermediate frequency was the essential feature. Whereas a triode valve would amplify quite well at 50 kc/s, it would become quite unstable at much higher frequencies. The early superhets thus frequency-changed the wanted frequencies down to a common (low) frequency at which the available valves could function effectively. This inherently led to difficulties; a low I.F. made the resultant set very sharp tuned, leaving only very truncated sidebands and cutting out all the higher speech – and more significantly – musical, frequencies.

When reasonable quality became important with improved L.F. circuits, and, in particular, loudspeakers, the superhet was increasingly dis-regarded. Other faults also led to its virtual demise. To return to our example, 200 kc/s gives an I.F. of 50 kc/s with a local oscillator set at 150 kc/s; also a local oscillator frequency of 250 kc/s would give an I.F. of (250 − 200) = 50 kc/s. Hence two quite separate oscillator dial settings could produce the same station at the loudspeaker. The situation was even worse than this, as harmonics of the oscillator could give rise to many other oscillator dial settings for the one incoming, wanted, transmission. Moreover, the possibility existed of other transmitted frequencies mixing with the local oscillator to generate the pre-set I.F; again in our example a transmitter working on 100 kc/s would mix with the 150 kc/s local oscillator to give 50 kc/s I.F.

The operation of one of these early superhets was fraught with difficulties. In addition, two extra valves were usually required: (single-valve autodyne mixer/oscillators were sometimes used). With early bright emitter valves this represented a substantial extra drain on the accu-mulator, perhaps more than 1 amp extra; also, more H.T. current was used, and, as well as the extra cost of the valves, there was the Marconi Royalty of 12/6 for each extra valveholder. This could add a substantial sum to the cost of the set. There was one other feature which helped to make the superhet very unpopular; it was necessary to mix the required transmission signal with the local oscillator in the frequency changer valve; when this was a triode, it was impossible to prevent the local oscillator frequency being strongly radiated from any external aerial attached to the set and acting as a transmitter. As this was not permitted, the only alternative was to use a highly inefficient, but domestically convenient 'frame aerial' – a large tuned coil which radiated poorly – and received just as poorly. Hence, unless the number of I.F. amplifier stages was increased, results also were poor. The catalogue of crimes set against the superhet in the early days was very great; nevertheless, as an elegant solution to a problem of great intractability, it had its devotees. It never entirely vanished from the scene in this country, indeed it had brief periods of considerable success, but always remained a set both expensive to purchase and to operate, and with many distinctly bad habits with which the owner had to learn to cope. Superhets were more common in America than here. The waveband congestion was substantially worse, valves were cheaper and there was no Marconi Royalty; and in any case more people could afford to have one. For many years the French also appeared to be addicted to superhets, though there were fewer listeners, and those who owned one were most probably well-off and could afford the extra costs. Only a superhet had the selectivity necessary to deal with the congestion.

A full account of this older form of superhet is given in 'The Practical Superhet Book' published by 'Amateur Wireless' towards the end of 1926.

The era of the Modern Superhet dawned only after 1927 when S.G. valves became available. These valves for the first time allowed good R.F. amplification and it is something of a paradox that the superhet in its old form was essentially an attempt to overcome this particular problem when only triodes were available. At about the same time, reliable 'coated filaments', and Indirectly Heated (I.H.) valves, came on the scene. Battery power for filament heating was no longer a severe limitation. The Marconi Royalty position was amended allowing a freer use of valves. In addition new mass-production manufacturing techniques for the associated I.F. transformers and other components permitted much more consistent characteristics.

With all these technical changes, the superhet came back into favour. The only real innovations were to use S.G.-I.F. valves and to shift the I.F. to higher frequencies; first to about 120 kc/s, later to 465 kc/s. At these higher frequencies it was much easier to obtain the 9 kc/s pass-band so essential for adequate quality of reception. The choice of I.F. was governed by the need to use a frequency well removed from normal broadcasting, otherwise a powerful transmitter close to the I.F. would inevitably 'break through' and be heard as a background to every required station. 465 kc/s equals 645m, and is just in the 'gap' between medium and long wave; few powerful stations transmit in this region. At these higher frequencies it was possible to design High Q tuned circuits and stagger-tune them to give a close approximation to the ideal pass-band.

References

1 The Wireless Constructor, February 1932, p. 232

Chapter 3
Broadcasting trends, 1924–34

Britain

At the start of 1924 the B.B.C. had in operation the 8 'main' 1.5 kW stations and 1 'relay' of 120 watts. By the end of the year, so rapid was the growth, all 9 main, 10 relay and the higher power experimental station 5XX at Chelmsford were in action. It was already clear that it would not be possible to extend coverage by an enhancement of the relay scheme as '. . . the nations of Europe were rapidly building up similar systems and inter- ference between stations even within their proper service areas was growing acute . . .'.[1] And yet the avowed policy of the B.B.C. was to provide an area of service to the listeners with crystal sets as great as was possible. At the end of 1924 there were about one million licence holders of whom 65% used crystal sets.[2] The experimental station 5XX was deemed to be the answer; it was experimental in that it broke entirely new ground, it was the most powerful transmitter in the world and indeed was designed by B.B.C. engineering staff. However much the B.B.C. might regard it as necessary in order to provide a wide area of reliable cover – and with the long wavelength of 1600m to avoid night-time fading – the concept was unproven. It was strongly opposed by the Post Office and other Government bodies. As early as one month after the transmitter started operation an editorial in Wireless World, whilst welcoming the new station, made the point that it, in turn, was interfering with the reception of Continental stations.[3] This 'swamping effect' was inevitable on the primitive single-tuned circuits of the time and was another factor slowing the advance to higher powers. It was an indicator of yet greater problems still to come. The new station, however, proved to be a marked success; re-siting (July 1925) to Daventry enabled it to provide a reliable service to 94% of the population (80% for crystal reception).

Eckersley had conceived the idea of 'Regional Broadcasting' as far back as 1924, even before 5XX had been tried out; the idea was to provide at five separate sites round the country 'twin wave' high power stations, each enabling a particular 'region' of the country to be served by two different

MAY 27TH. 1925.   THE WIRELESS WORLD   ADVERTISEMENTS.   3

programmes radiated on different frequencies from the same site. It thus had as one of its aims the desire to provide alternative programmes. In this scheme Eckersley was strongly supported by Reith who felt that '. . . a loss of interest in the service as well as loss of pride of place in world broadcasting'[4] would occur unless some such scheme were to be introduced. According to Reith the public wanted to hear other stations and the Company would have to provide them if it wished to retain control of broadcasting. He foresaw the Regional Scheme as hastening the demise of the crystal set ('. . . the crystal receiver . . . now being put to a severe test.' '. . . crystal set joining company with the hansom cab and the gingham umbrella'),[5&6] and other unselective obsolete equipment and so it could only be introduced slowly. Just how slowly official opposition was overcome can be judged by the apparently grudging permission, not granted until 1926, to allow a further 'experimental' station at Daventry. This, as 5GB, began to transmit on medium wave with a power of 30kW on 21 August 1927[7] and the former 'main' station 5IT of 1.5 kW was closed.

The new high power stations caused listeners many problems; older equipment for the most part simply was not able to separate such powerful transmitters. Even although one was on the long wave band and the other on medium wave, they had never been intended for reception under such conditions. For many, the 'sport' of long distance Continental reception became impossible, swamped by the radiated power; for still others, reception was less satisfactory than with 5IT. Nevertheless, after the dust had settled, the scheme was adjudged to be successful and the B.B.C. learned much that enabled it to help other listeners when later high-power stations opened.

After considerable speculation, the B.B.C. was able to make a start on a second twin-wave station at Brookmans Park, fifteen miles to the North of London, in July 1928. The station opened on 21 October 1929. 2LO closed. With the experience of Birmingham behind it the B.B.C. was able to advise listeners by means of pamphlets how to cope with the changed conditions and very much less disruption occurred. Moorside Edge, the North Regional, followed on 12 July 1931; Scottish Regional at Westerglen in September 1932; West (and Welsh) Regional in May 1933 (April 1933 according to Ashbridge).[8] Notably, West and London had to share a synchronized wavelength of 261.6m as the B.B.C. had been allocated only a limited number of frequencies. Consequently with these changes, many older stations were altered; in Scotland, Aberdeen became a relay, Edinburgh, Glasgow and Dundee were closed. In the West, Cardiff and Swansea closed, Plymouth and Bournemouth, however, were retained as there were special difficulties in these areas. Wireless World was able to report a 'Programme Innovation';[9] from 8 October there was to be a continuous alternative programme radiated from 10.45 a.m. onwards. Noel Ashbridge, B.B.C. Chief Engineer after Eckersley resigned, indicated that some 98% of the entire population of the country had one 'National' programme, 85% had two, National and Regional, at good strength.[10]

MAY 27th, 1925. Wireless World 501

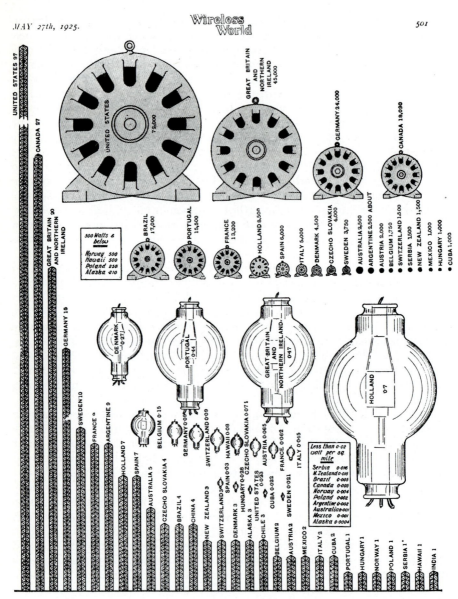

Height of masts = number of stations; size of generators = output; size of valves = watts per square mile.

Further developments took place subsequently; Droitwich at 150kW replaced Daventry 5XX in October 1934; London, North and West Nationals could all then be closed, freeing two of the B.B.C's allocated frequencies and allowing better reception to be provided in the North of England and in the far North of Scotland. All through the implementations of the scheme B.B.C. wavelengths were reshuffled many times as plan succeeded plan in an effort to clear the chaos in the European airwaves.

Over the decade 1924 to 1934 therefore, the whole British transmitting scene underwent continuous change; going from a series of 'main' and 'relay' stations of 1.5kW and 120 watts to the 'Regional' – of 30kW – and 'National' a monster of 150kW. Such changes of themselves did much to require design changes in British receivers. Others were forced by the equally dramatic changes in European broadcasting.

Continental Europe

During the 1924–34 era, Continental broadcasting was spreading and increasing in power. At the start of 1924, Wireless World listed seventeen stations in mainland Europe most transmitting irregularly and infrequently;[11] by the end, there were twenty-five regular and a further fourteen testing.[12] The rate of growth is well illustrated by a pictogram.[13] This shows the B.B.C. to have twenty stations with an aggregate output of 45kW, whilst the rest of Europe had already established about seventy stations with a total radiated power of 93kW. With the exception of 5XX and, curiously, Portugal, all were of comparatively low power, averaging 1.5kW. Just two years later there were two hundred and thirty European transmitters with an aggregate of 550kW, an average of 2.4kW;[14] thus, not only were numbers of transmitters growing, but the move to higher power was showing. Even low-power stations can, especially after dark, have very considerable range and cause interference with distant stations. At this time there was no attempt to 'order' the allocation of frequencies; each new transmitter appropriating whatever frequency it fancied: a recipe for chaos. It is interesting to note that this scene can also be found in the U.S.A. and indeed persisted there long after co-operative efforts in Europe had led to some improvements. This fact had a considerable influence on design-differences apparent in receivers from opposite sides of the Atlantic.

The frequency bands utilised by these swelling numbers of stations were restricted to 103 to 470 kc/s (2900 to 640m) long wave and 510 to 1690 kc/s (590 to 178m) medium wave. These divisions were not fixed, nor rigidly adhered to, but in practice these were the broadcasting bands in use. Since each transmitter required a 10 kc/s (later 9 kc/s) bandwidth to accept reasonable modulation frequencies, then at the very best there could only be $\frac{470 - 103}{10} = 37$ unique long wave frequencies and $\frac{1690 - 510}{10} = 118$ on medium wave, a total of 155; with 230 stations already operating. Stations geographically far apart could share frequencies without unacceptable mutual interference in daylight hours at least; so there could be more stations than frequency channels, but the figure of 230 was clearly too many. Moreover, the situation was worse than even this figure suggests; broadcasting had already encroached on regions of the spectrum reserved for 'essential' Government services, shipping and distress signalling. Also, even with selectivity far better than the norm for this time, a transmitter close to the receiver would appear to 'spread' over 30 or even 50 kc/s. Early receivers had very poor selectivity; nothing more was expected of them other than reception of the nearest transmitter. Difficulties were enhanced at night with the reflective ionised layers (Heaviside layer) giving vastly greater range to even low power radiators; also 'fading' became a particular problem (interference between ground and reflected ray from a relatively close transmitter).

Considerations such as these soon led the B.B.C., through Eckersley, to a realization that more was required to develop broadcasting than a proliferation of low-power relays on no-longer-available frequencies. As a temporary palliative, and in order to clear some wavelengths for the new main transmitters, low power relays were worked synchronously on a common wavelength. At least some of the interference that such a practice should give rise to was overcome by accurately tuning each by means of a tuning-fork frequency standard. Very remarkable results could be achieved by maintaining this temperature-controlled frequency standard. Deviations of no more than a few parts per million were possible, and the technical expertise built up was invaluable for later requirements.

Bad as the situation was in Europe, it was many times worse in America; a two year lead in broadcasting had allowed so many stations to be in operation that reception conditions were very poor. Much more important, however, the unbridled commercialism of American radio ensured the maximum number of transmitters all '. . . piling kilowatt upon kilowatt'.[15] Special measures were necessary, in receivers, to separate required stations. The American answer, encouraged by a lack of Marconi Royalty payments and the mass production of cheap valves, was to have multiple, tuned R.F. stages in multi-valve receivers.

In Europe it was recognized that improved reception conditions were attainable provided the international free-for-all on the wavebands could be called to order and transmitters organized into equally spaced frequency slots. The B.B.C. was in the forefront in attempting to secure some agreement and eventually an international conference was convened in Geneva, opening on 25 March 1926 Wireless World wrote in anticipation 'To-morrow week, Thursday, 25 March, must be regarded as an important date in the annals of European Broadcasting . . . representatives of all existing or projected broadcasting organizations in Europe will confer at the Palais des Nations, Geneva, for the purpose of redistributing broadcast wavelengths. A reduction in the number of existing stations is included in the recommendations . . . accompanied by an increase in power'.[16] This was followed with a summary of the conclusions. 'The outcome . . . is that the international experts are uniformly agreed on the B.B.C. idea of building up broadcast schemes along the lines of higher power and the elimination of most of the low-power stations. The present system of building without regard for the limited capacity of the waveband available cannot, however, be changed at a moment's notice without . . . dislocation and inconvenience. . . By the duplication of wavelengths in suitably chosen zonal areas a good deal of . . . interference will be overcome, and attention can then be turned to the more delicate task of limiting the number of stations that is considered adequate for each of the broadcasting nations'.[17] Further progress was reported after 'Europe's Wavelength Problems'. 'The Committee of the International Wireless Union . . . met on 5 and 6 July in Paris'[18] and approved the Report of the President of the Technical Commission M. Raymond Braillard, regarding the redistribution of wavelengths between 200 and 600m (to which the medium wave was to be confined). Further plans for the longer wavelengths were said to be in preparation.

Full details of the allocation were published in Wireless World where the article refers to 'The most important co-operative move yet made by the broadcasting stations of Europe will take place on 15 September . . .'.[19] Broadly, the plan divided the available frequency range into 10 kc/s 'slots'; 589m to 201m is 511 kc/s to 1491 kc/s, a range of 980 kc/s thus allowing for ninety-nine frequency slots. Eighty-three of these were exclusively allocated to individual transmitters and sixteen to 'common wave', shared by ninety-six lower power transmitters. Within these allocations, the B.B.C. had seven exclusive wavelengths; Glasgow, Manchester, London, Cardiff, Belfast, Newcastle and Bournemouth, a further one exclusive wavelength which it shared between Aberdeen and Birmingham, two neighbouring slots for Leeds and Bradford, both of which were also used by five or six Continentals, and one common user for Dundee, Edinburgh, Hull, Liverpool, Nottingham, Plymouth, Sheffield, Stoke and Swansea: a total of eleven slots for its twenty transmitters. No mention was at this time made of the long wave, but 5XX continued to function. The highest power registered in the medium wave was of 5kW, but overall did not average as much as 1kW.

The American experience is highlighted in an article 'Summer Radio in America'. It refers to five hundred and sixty-six current broadcasting stations, six hundred applications for new ones; many are of low power, but the principal stations are increasing in power up to 50kW; in fact a growth in total of 200kW in one year, leading to 'chaotic results'.[20]

In Europe, the wavelength plan was due to come into effect on 15 September 1926, but the new order was delayed for a few weeks[21&22] seemingly to allow calibrated wave meters to be sent out.[23] The plan came into effect on 21 November 1926. 'Bravo Geneva!' 'On the whole, the system is giving satisfaction . . . a certain amount of trouble was experienced . . . but every day now sees an improvement'.[24] A new complete list of wavelengths is given in Wireless World;[25] it is interesting to note that it gives twenty-one long wave stations also, and a few between medium and long wave bands. By far the most powerful is Daventry, 5XX at 25kW on 1600m; with Moscow, 12kW on 1450m and Berlin, Frankfurt and Hamburg all at 10kW, the only others in double figures. A comment in Wireless World is interesting in this connection, it notes that one possible flaw in the Geneva scheme is that it took no account of transmitter power.[26] An article of 24 November 1926, contains the comment '. . . the wavelength allotted . . . will probably remain . . . for a considerable period',[27] but within two weeks the deliberations had been '. . . resumed at Geneva'; in the UK common wave working for all the relay stations had resulted in pandemonium. There is also a list of further revised wavelengths for British stations in which many of the relays are re-located and a comment is made that the B.B.C. may press ahead with the Regional Scheme so that some relays could disappear '. . . next year'.[28] It also suggests 10kW, favourite amongst the larger Continental transmitters, may be used. By 19 January 1927, it is reported that things were settling down, but the new menace of high-power stations was beginning to trouble the writer.[29] Shortly after there were yet again some shifts of B.B.C. frequencies.[30] At the same time the start of the Regional Scheme is foreshadowed by the announcement that 'Daventry Junior' – the future 5GB – would start medium wave tests '. . . in three or four weeks'.[31] The Union Internationale de Radiophonie is reported to be starting to tackle the long wave spectrum at a conference at Brussels.[32]

Attempts to put the long wave stations on a satisfactory basis seem to have been placed largely in Eckersley's hands, as he organized the co-operative operation run at night to find experimentally what could best be done.[33] Considerable numbers of frequency revisions took place through-out the next year, culminating in the Brussels Plan which came into operation on 13 January 1929; this was not an entirely new scheme, but more in the nature of a reshuffle along with a narrowing of the 'slots' in some cases, to 9 kc/s width.

Amongst all this instability, arrangements were under way for another International Conference, this time in the U.S.A., at Washington. It was opened on 4 October 1927 by the President of the U.S.A. The U.S. Secretary of Commerce, Hoover, was elected as Chairman and P. P. Eckersley represented the B.B.C.[34&35]

The resultant International Radio-Telegraphy Convention was signed in Washington on 25 November 1927, by eighty nations and was effective from 1 January 1929. The conference was very wide-ranging in its discussions, and agreed amongst other things, despite very stiff American opposition, to recognize long wave band as well as medium wave, for broadcasting, thus allowing 5XX to continue. It was most probably American commercial interests which opposed long wave, as it was not in use in the U.S.A.

The ever-growing difficulties in Europe pointed to the need for yet another attempt to produce order; reports that the '. . . Geneva "Scheme" . . . is beginning to flounder'[36] and 'More and more . . . B.B.C. stations are being heterodyned . . .'[37] were common. A wry comment was that the new 'game' was to see how many stations could be received – all at once![38] These experiences are paralleled by similar reports from the Continent. One fundamental difficulty after the Geneva and Brussels Plans was that there was no power to enforce the resolutions, this was recognized and allowed for at the Washington Conference.[39] As a follow up, the

International Wireless Conference of twenty-six European Governments was held in Prague from early in April 1929. The agreement provided for fifteen L.W. stations and one hundred and twenty-three on the M.W; frequency differences were to be 9 kc/s. The B.B.C. had 5XX on long wave as well as eight exclusive M.W. slots. These allocations could now be enforced since Governments had agreed to the provisions and the Bureau Internationale de Radiophonie in Brussels was empowered to police the observance of the agreements. Doubts were rife as to the likelihood of success; an article in Wireless World is headed 'Is the Prague Plan Sound' and after a detailed analysis of possible diffi- culties makes the telling point '. . . the ardent desire shown by certain foreign States to possess super-power transmitters' (Germany, Russia). The article concludes 'The Prague Plan must surely be a provisional measure only'.[40] The new arrangements came into operation on 30 June 1929 and the B.B.C. again, now without Eckersley who resigned earlier in the month, assisted in experiments to check the operation. Early reports were encouraging, but as it was midsummer this was perhaps not so surprising.

As well as the steady move to higher power, the political situation in Europe led to more use of Propaganda Broadcasting, moving Wireless World to an editorial entitled 'The Battle of the Giants',[41] and another 'The Thirst for Power'.[42] A conference at the Hague had set a limit of 100kW for broadcasting. According to the editorial this 'limit' was seen by some Governments, particularly by Germany, as the goal to be aimed at for all stations. Frequent references are made to the rush for power in the following months: 'One by one the Continental transmitters are increasing their power'.[43] 'Broadcasting On 200 Kilowatts. The New Prague Trans- mitter . . .'.[44] The growth in both power and numbers of stations is shown most clearly in graphs illustrating an article by Noel Ashbridge.[45] Very briefly these show a steady increase in the number of broadcasting stations of Europe, reaching 277 by the start of 1933, with a sharply rising aggregate power reaching 4600kW – an average of 16.5 kW/station – and a maximum for any station of 150kW.

A further attempt to bring the situation under control was introduced with the Lucerne Plan of 1933, following ground work at Lugano and Madrid in 1931 and 1932. Briefly, the Lucerne Plan again maintained a 9 kc/s separation; but it also sought to control power and as expressed in a Wireless World article 'No station will be permitted to change its power or its wave without . . . the approval of all concerned'.[46] This plan in its turn was put into operation on 14 January 1934. The almost complete lack of comment over the next several months presumably speaks for itself. In any case, other matters were beginning to obtrude 'Politics are the curse of broadcasting to-day'.[47]

During the 1924-34 decade European Broadcasting was entirely trans- formed, from a few low power transmitters scattered in location and frequency, to the continuous band of stations 9 kc/s apart, many at high power, and all vying for listeners, not necessarily within their own geographical region.

References

1 A. Briggs, The History of Broadcasting in the United Kingdom, Vol. II, p. 294 (Eckersley, Regional Scheme Report, 20 June 1927)
2 Amateur Wireless, 14 February 1925, p. 320
3 Wireless World, 20 August 1924, editorial
4 A. Briggs, The History of Broadcasting in the United Kingdom, Vol. II, p. 296
5 Wireless World, 31 August 1927, editorial
6 Wireless World, 24 August 1927, p. 249
7 Wireless World, 17 August 1927, p. 217
8 Wireless World, 13 July 1934, pp. 24ff
9 Wireless World, 14 September 1934, p. 233
10 Wireless World, 13 July 1934, pp. 24ff
11 Wireless World, 9 January 1924, p. 478
12 Wireless World, 7 January 1925, p. 496
13 Wireless World, 27 May 1925, p. 501
14 Wireless World, 15 June 1927, p. 760
15 Wireless World, 18 August 1926, pp. 229ff
16 Wireless World, 17 March 1926, p. 409
17 Wireless World, 7 April 1926, pp. 533ff
18 Wireless World, 21 July 1926, p. 91
19 Wireless World, 1 September 1926, p. 311
20 Wireless World, 18 August 1926, pp. 229ff
21 Wireless World, 8 September 1926, p. 365
22 Wireless World, 13 October 1926, p. 529
23 Wireless World, 3 November 1926, p. 602 and p. 618
24 Wireless World, 24 November 1926, p. 719
25 Wireless World, 24 November 1926, p. 700
26 Wireless World, 17 November 1926, p. 686
27 Wireless World, 24 November 1926, p. 700
28 Wireless World, 8 December 1926, p. 781
29 Wireless World, 19 January 1927, p. 85
30 Wireless World, 26 January 1927, p. 120
31 Wireless World, 26 January 1927, p. 119
32 Wireless World, 26 January 1927, p. 119
33 Wireless World, 9 February 1927, p. 179
34 Wireless World, 5 October 1927, p. 495
35 Wireless World, 12 October 1927, p. 518
36 Wireless World, 5 October 1927, p. 495
37 Wireless World, 5 December 1928, p. 773
38 Wireless World, 20 March 1929, p. 312
39 Wireless World, 4 May 1927, p. 568
40 Wireless World, 19 June 1929, pp. 638ff
41 Wireless World, 2 October 1929, editorial
42 Wireless World, 27 November 1929, editorial
43 Wireless World, 18 November 1931, p. 584
44 Wireless World, 25 November 1931, p. 619
45 Wireless World, 19 August 1932, pp. 146ff
46 Wireless World, 26 May 1933, p. 373
47 Wireless World, 30 March 1934, p. 216

Chapter 4

British radio valves, 1924–34

During the 1924–34 period, fundamental changes took place in radio valves; manufacturing methods, types and designs were all completely transformed. During the war, in particular, the valve had emerged from its hand-made, laboratory status and had transferred to large scale, production-line manufacture. It had, however, scarcely changed in any essential way. By 1922 the Marconi Osram Valve Co. (M.O.V.) had brought out LT1 and LT3 types with Thoriated Tungsten filaments which were much more economical in filament heating power; later DE3 and from British Thompson Houston (B.T.H) B5 types which had very fine filaments requiring only 60 m/A at about 3V. There was also a realization that different valve characteristics were more suitable for different positions in a receiver, e.g. a power output valve was more efficient if it had a reasonably low anode impedance; there was competition to produce detector valves which were not a prolific source of 'microphonic' noises; and there were one or two odd 'special' types – perhaps with an extra grid incorporated – but these were in many cases only partly understood and saw little commercial use.

   In truth, even in 1925, all the multitude of types from great numbers of manufacturers were all very similar triodes, many of which were, apart from filament characteristics, in many ways the same as those available ten years earlier, although attempts had been made to develop power types. Valves, especially those from the large manufacturers, were very expensive; a single dull-emitter, which, with its characteristic top-pip, silver 'gettering' and resplendent on its bright nickelled base, was clearly intended to be on full display, would cost more than a complete crystal receiver, headphones, aerial and earth equipment all included. Two, less obvious but nonetheless powerful influences on progress of valve design were (1) the Marconi Royalty, (2) the requirement for all apparatus used in this country to be of British manufacture, the 'B.B.C' stamp era. The first was a charge (initially 12/6) levied by the Marconi Company on each valve holder in consequence of the patents they held. Because of this extra charge – on top of an already high price – British wireless tended to be

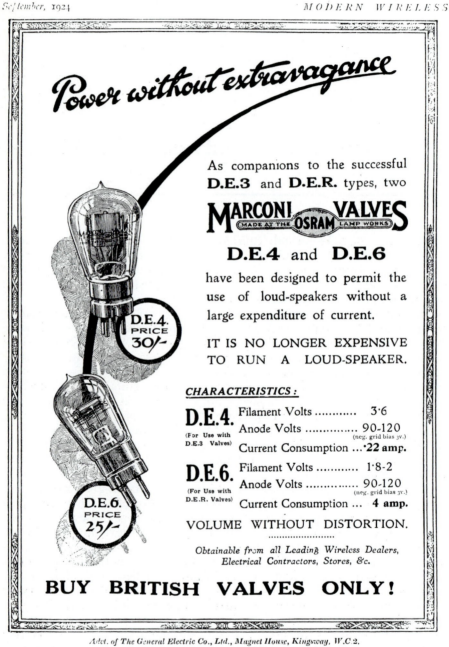

Power without extravagance

As companions to the successful **D.E.3** and **D.E.R.** types, two

MARCONI OSRAM **VALVES**
MADE AT THE LAMP WORKS

D.E.4 and **D.E.6**

have been designed to permit the use of loud-speakers without a large expenditure of current.

IT IS NO LONGER EXPENSIVE TO RUN A LOUD-SPEAKER.

CHARACTERISTICS :

D.E.4.
(For Use with D.E.3 Valves)

Filament Volts	3·6
Anode Volts	90-120 (neg. grid bias 3v.)
Current Consumption ...	·22 amp.

D.E.6.
(For Use with D.E.R. Valves)

Filament Volts	1·8-2
Anode Volts	90-120 (neg. grid bias 3v.)
Current Consumption ...	4 amp.

VOLUME WITHOUT DISTORTION.

Obtainable from all Leading Wireless Dealers, Electrical Contractors, Stores, &c.

D.E.4. PRICE **30/-**

D.E.6. PRICE **25/-**

BUY BRITISH VALVES ONLY!

Advt. of The General Electric Co., Ltd., Magnet House, Kingsway, W.C.2.
In replying to advertisers, us COUPON on last page

     ADVERTISEMENTS.

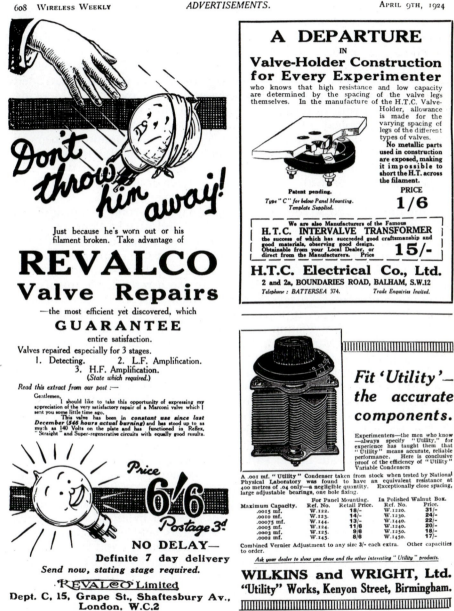

very 'economical' with valves. Apart from maintaining high prices, this very fact also encouraged attempts to develop new types which, by their very efficiency, could reduce the number of stages required. The 'B.B.C' restrictions also ensured that the native manufacturers were insulated, in their early growing days, from external competition. This also helped to maintain high prices; curiously, however, these twin pressures provided the spur – and the income – to finance a very active and forward-looking valve-manufacturing industry which ensured that new techniques were discovered, or applied – for by no means all were of native origin – with remarkable rapidity. There were a number of powerful manufacturers all in active competition with each other, but it was long enough before prices began to fall, a subject for much bitter comment throughout the 1920s, but perhaps it can be justly said that British valve design and manufacture was at least the equal of any in the world.

By the start of 1926, so routine had valve-making become that in addition to the large manufacturers, the success of the industry had encouraged a great variety of large and small independent firms to sprout. Many of these had grown out of the 'valve repair' firms; with a valve costing as much as a week's wage, there was great incentive to replace the filament of a burnt-out valve for half-price. During 1926 there was a remarkable upsurge in the number of manufacturers, some of which were really of American origin; amongst the types appearing in the pages of Wireless World for the year were: Amplion, Benjamin (Shortpath), B.S.A. (Standard), Cleartron, Lustrolux, Nelson (Multivalve), Neutron, Octron, Professor Low, S.T. (Scott-Taggart) – most of these appeared within a few weeks of each other – and most disappeared shortly afterwards. Besides these aberrations, major developments were taking place with very far reaching consequences.

The first of these was the development of valves capable of having their electron-emitting surfaces heated by raw A.C. The background to this is tangled, but reference can be made to British patent No. 213,605 taken out by N. V. Philips describing in all essential detail the solution eventually adopted, of a separate heater cathode assembly.[1] The electron-emitting surface had to be capable of producing a copious stream of electrons at temperatures far below that of the customary directly heated filaments. Materials – rare earth oxides – were available for this purpose, but it was very difficult to ensure that these oxides adhered firmly to the cathode; possibly the Mullard P.M.4 was the first in this field for directly heated coated filaments in this country.[2] Cossor also had the W1 and W2 available, but they were in essence the original bright emitter P1 and P2 valves with a normal Tungsten filament 'triple coated', they seem to have been not very satisfactory and might have been no more than a stop-gap to keep up with their arch-rivals.

Although the patent mentioned above was taken out in 1924, it was not until 1927 that M.O.V appeared ready to launch their first 'indirectly heated' (I.H) valve, the KL1,[3] priced at 30/-. The 'coating' process for battery valves appears to have become stabilised by then as they had become quite common; the Cossor 'point one' series is advertised 'Once again, Cossor has blazed a trail in Valve design, first in 1922 with an arched

iv. ADVERTISEMENTS.   THE WIRELESS WORLD AND RADIO REVIEW   FEBRUARY 2ND, 1927.

POWER DIRECT FROM A.C. MAINS

WIRELESS WITHOUT BATTERIES

THE LATEST MARCONI valve achievement enables you to obtain the necessary power for operating a radio set through the electric light socket direct from A.C. Mains, thus entirely dispensing with accumulators or batteries

MARCONI TYPE K.L.1 employs a new principle in radio valve design. The electrons are not emitted from the filament but from a separate cathode heated by thermal radiation.

MARCONI TYPE K.L.1 is a general purpose valve operated through a special transformer which supplies up to 8 amperes at 4 volts for the filament, 150-150 volts for the H.T., and 5 volts 2 amperes for a U5 Valve used as a rectifier for the H.T. Supply. Although adaptable to almost any existing receiver, a special circuit is required, particulars of which are obtainable on request

from your usual radio dealer or direct from
THE MARCONIPHONE COMPANY LTD.,

Registered Office :—   *Head Office :—*
Marconi House, Strand, London, W.C.2.   210-212, Tottenham Court Rd., London, W.1.

MARCONI—THE GREATEST NAME IN RADIO

MARCONI
VALVE TYPE K.L.1

APPROXIMATE DATA :—

Fil. volts	-	3·5	*Fil. Current*	-	*2 amperes*
Anode volts	-	100	*Amp. factor*	-	*7·5*
Impedance	-	5,500 ohms	*Normal slope*	-	*1·30 ma/volt*

M.P.O.863C.

WRITE FOR SPECIAL CIRCUIT DETAILS

Printed for the Publishers, ILIFFE & SONS LTD., Dorset House, Tudor Street, London, E.C.4, by The Cornwall Press Ltd., Paris Garden, Stamford Street, London, S.E.1.

Colonial and Foreign Agents:
UNITED STATES –The International News Co., 83-85, Duane Street, New York.   FRANCE—W. H. Smith & Son, 248, Rue Rivoli, Paris; Hachette et Cie, Rue Reaumur, Paris. BELGIUM—W. H. Smith & Son, 78, Marche aux Herbes, Brussels.   INDIA—A. H. Wheeler & Co., Bombay, Allahabad and Calcutta.   SOUTH AFRICA—Central News Agency, Ltd. AUSTRALIA—Gordon & Gotch, Ltd., Melbourne (Victoria), Sydney (N.S.W.), Brisbane (Queensland), Adelaide (S.A.), Perth (W.A.) and Launceston (Tasmania). CANADA—The American News Co., Ltd., Toronto, Winnipeg, Vancouver, Montreal, Ottawa, St. John, Halifax, Hamilton; Gordon & Gotch, Ltd., Toronto; Imperial News Co., Toronto; Montreal, Winnipeg, Vancouver, Victoria.   NEW ZEALAND—Gordon & Gotch, Ltd., Wellington, Auckland, Christchurch and Dunedin.

filament operating within an electron-retaining hood-shape Anode system – first in 1924 with a triple-coated filament producing a prolific electron stream at a phenomenally low temperature . . .'.[4] However this may be, even when Osram introduced the KL1, the valves were obviously still a rather unsatisfactory product; it is noteworthy that even an M.O.V. advertisement refers only to 'approximate data-'[5] for the characteristics, and an article on them in Wireless World reports some difficulty in making stable measurements. The same article, incidentally, refers to American experience and includes the phrase '. . . has not attained a wide popularity'[6] with reference to coated emitters. M.O.V. pressed on with the KL1, however; another advertisement describes a design for a complete K1 receiver using I.H. valves with the encouraging statement that it was 'Easily assembled from the very complete details and full size wiring plan given in the K1 Constructional Book'.[7&8] Transformers for supplying the required heating current appeared with remarkable rapidity[9] and some idea of the interest stirred up by the new valve can be judged by a report of the Proceedings of the Kensington Radio Society held in February where the new valve was fully explained and described by a Mr Harwood of the Marconiphone Co. Ltd.[10] It is, however, probable that all was not well with the new valve; in spite of some further publicity up until May, the valve was, significantly, not listed in the Wireless World 'Valve Supplement'[11] and, perhaps an even more telling absence, it did not even appear in the 'additions'.[12] An article on the KL1 occurs in Wireless World.[13] Here is described the 'white hot' tungsten heater, the nickel cathode with its emitting coating and comparison is made with the venerable LS2 bright emitter; apart from the ability to use raw A.C. for heating, very little advance in the characteristics of the valve are discernible. It continued to appear in various M.O.V. publications, was listed in the 1929 Wireless World Valve Data Sheet, but had vanished from the scene by 1930.

These valves (a KL2 was also briefly in evidence) never seem to have had wide usage and a suspicion lingers that they were not entirely satisfactory; it is possible that the writer of a rather sour editorial in Wireless World had them in mind when he wrote 'We cannot overlook the fact that at the Radio Show of 1927 certain valves were exhibited which never actually came on the market, . . . and those which were available often showed a distressing deviation from type'.[14] Whether this be the case or not, M.O.V. were hedging their bets, and following Continental practice produced in time for the 1928 Show a series of valves with short, thick filaments taking 0.8 amp at 0.8 volt which were intended to be run on raw A.C. For use with these directly heated A.C. valves an indirectly heated type had to be used as a detector. These valves, too, seem to have had a chequered career, they reappeared in 1930, now with the addition of a detector, D8, taking 1.6 amp at 0.8 volt, but all vanished by 1931. One set appeared using 0.8 series valves; '. . . two-valve all-electric receiver. The detector valve, a D8, is followed by a super-power valve P625A'.[15] A similar duality is noted in American practice where the UX226 uses raw A.C. on its filament and the UY227 has a heater/cathode assembly.[16]

A Valve which operates from A.C. Mains Supply

REVOLUTIONARY BRITISH INVENTION
The New Osram **K.L.1.** Valve

PRICE
30/-

"'TONE,' we've made 1927 a wonder year in wireless," said 'POWER' proudly. "Our new K.L.1 OSRAM VALVE is something entirely new, and a revolutionary advance in wireless."

"True enough," replied 'TONE.' "Now listeners have a valve which operates from A.C. mains supply with a simple transformer. There is no hum, no danger, wonderful amplification, and no L.T. batteries to run down."

"Once again we show the way to Better Wireless Reception and register yet another triumph for *BRITISH INVENTION*."

"Every listener with A.C. mains supply should write for particulars to The General Electric Co. Ltd., Magnet House, Kingsway, London, W.C.2."

'TONE' & 'POWER.'

Osram Valves
for TONE & POWER

The G.E.C.- your guarantee

Advt. of The General Electric Co. Ltd., Magnet House, Kingsway, London, W.C.2.

B3   *Advertisements for " The Wireless World " are only accepted from firms we believe to be thoroughly reliable.*

VALVES OF CHARACTER
FOR OPERATION OFF THE ELECTRIC LIGHT

Don't engage a Valve
without a good character

"Cosmos-Met-Vick" A.C. Valves are each supplied with a written character, the details of which are in close accord with the actual inherent character of the valve.

The new reduced prices are comparable with those of ordinary battery valves and will greatly assist all who are converting their sets from battery working to operation from the electric light mains.

The A.C./G (Green Spot) Valve can be used for any stage except the last.   It has a very high amplification factor of 35 with an impedance of only 17,500 ohms.   It is suitable as a Detector and for all forms of coupling.

Used by Mr. N. P. Vincer-Minter in his A.C.2 and A.C.3 (*Wireless World, August 22nd and September 5th.*)

The A.C./R (Red Spot) valve has been designed specially for the Loud Speaker Stage. It has a very high mutual conductance, and amplification factor of 10 with an impedance of 2,500 ohms at 180 volts H.T.

It will give twice the output for the same input of any battery operated valve on the market.

EVERY USER IS ENTHUSIASTIC ABOUT

"COSMOS" A.C. VALVES

PRICES NOW
REDUCED
TO

15/-   17/6
(GREEN SPOT)   (RED SPOT)

METRO-VICK SUPPLIES
LIMITED
Proprietors:
Metropolitan-Vickers Electrical Co. Ltd.

155 Charing Cross Rd.,
LONDON, W.C.2.

See us
at
OLYMPIA
STANDS
32
AND
41

R
V119

A3   *Advertisements for " The Wireless World " are only accepted from firms we believe to be thoroughly reliable.*

Other manufacturers were quick to produce I.H. valves of their own design: notably Metro-Vick with their "Cosmos" AC/R and AC/G. These two, which had very much better characteristics than the M.O.V. types, made a substantial impact and set the pattern for future developments, although their unusual base was not perpetuated. Cossor followed some months behind with some rather curiously shaped laboratory type valves. Mullard, too entered the field but much later. Some idea of the rapidity of growth of the indirectly heated valve market may be judged from the Wireless World Valve Data Charts:

Table 1

Year	No. of valve types	No. of I.H.
1927	337	0
1928	180	13
1929	253	13
1930	338	46
1931	518	113

NOTE: In Wireless World several types are incorrectly described as 'directly heated'.[17]

The future, in general, clearly lay with the I.H. mains valve; except for special purposes, – country districts with no mains supply, and for portable sets, – the long standing standard filament-type valve was overshadowed by its mains-operated rival. One symptom of this was that within a year or two almost all except the 2 volt filament types had vanished, and a great clear-out of obsolete specimens took place. About this time, too, valve prices dropped rapidly and this hastened the demise of the small, independent, maker. It is probable that the technique of manufacturing I.H. types was just too much for their resources, and the low prices could not have allowed much of a profit margin anyway.

The other major valve revolution was brought about by the development of the screen grid valve. First used exclusively for radio frequency amplification, later, with an additional grid, it became the pentode with wide application in all stages in the receiver.

The first, very complete, account of the properties of S.G. valves was given by A. W. Hull and N. H. Williams of the Research Laboratory of the (American) General Electric Company, in a paper dated 30 December 1925 and published in the 'Physical Review', Vol. 27, p. 432: April 1926. An adjoining paper by Hull is on the 'Measurement of High Frequency Amplification', making use of the new design. The significance of these new discoveries was very great; for the first time it was possible to produce a straightforward, stable, high frequency amplifier. The only previous relatively successful method using a triode had required a difficult 'neutralizing' procedure to overcome the positive feedback problems created by the 'Miller Effect'. Neutralized triode R.F. amplifiers had been considerably developed, but never found much favour with set manufacturers. Unless they were peculiarly well designed, performance was poor and long term stability uncertain; stable variable tuning circuits were quite difficult to make. The new valve could give stage gains of around 100 compared to perhaps one-tenth of that for even a carefully neutralized triode.

The way forward in this country seems to have been pioneered by Captain H. J. Round, of Marconi, who had made notable contributions to wireless in war and peace. Wireless World gives details of two British patents issued to Round; No. 275,335 – application date 5 May 1926 – for a double-ended design in all essentials identical with the valve which finally emerged as the M.O.V. S625, No. 279,171 – application date 22 July 1926 – for a modified design, only the anode lead brought through the top of the bulb.[18&19] This latter became the standard form in this country for many years. Interestingly, Hull's original design had the inner (control) grid connection made in this manner and this was subsequently usually adopted. Hull's paper appeared in April 1926; Round's patent application appears at more or less the same time. Hull was the inventor of British patent No. 230,011 and Dewey Theodore Simonds, was the inventor of British patent of addition No. 255,441, both patents preceded Round's. Marconi's produced the FE1 valve in 1920, developed by Round as a screen grid; but the Company failed to exploit its capabilities and used it only in peculiar reflex circuits. Round's S625 was a 1926 revival of his earlier ideas. Round published a book 'The Shielded Four-Electrode Valve Theory and Practice'.[20]

'The New Screened Valve'[21] was the title of an article in Wireless World by N. W. McLachlan; the form of the valve as shown there is exactly as in Round's original patent (5 May) and it is operated under the conditions specified therein. The form of the valve is quite different from that advocated by Hull and, in fact, requires an external screen – which is specified in the patent – to achieve the enhanced screening effect shown by Hull to be necessary and incorporated by him in his design. The 'Round' valve gives every appearance of having been put together from parts already to hand; the filament and first grid are those of the then current DE5 valve, but mounted horizontally; the mesh screen and circular plate are new and mounted on a separate 'foot' at the other end of the cylindrical glass valve; such a design must have made it virtually impossible to produce consistency in characteristics. An accompanying article by W. James shows how to incorporate the new valve, called – as by Hull – a 'shielded' valve, in one of Wireless World's recent receivers, the 'New Everyman Four'. On two separate occasions James draws attention to the fact that '... valves vary, some having a very much higher anode impedance ...', just what would be expected from slight differences in spacing between the two halves. The design suggested quotes a figure of 35 for the amplification of the complete stage compared '... with a pure radio frequency magnification of 40 for a standard "Everyman Four" coil, which is increased by

THE NEW EVERYMAN FOUR

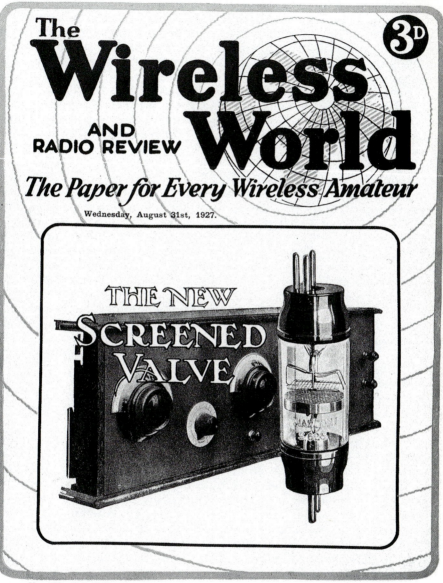

The

Wireless

AND
RADIO REVIEW

World

3D

The Paper for Every Wireless Amateur

Wednesday, August 31st, 1927.

THE NEW
SCREENED
VALVE

No. 418. Vol. XXI. No. 9.

... the regenerative contribution obtained when the balancing condenser is carefully adjusted'.[22] No comment is made about the performance of the receiver; the following issue continues McLachlan's article, and shows the effect on selectivity of the new valve as compared with a neutralized triode (horizontal scale is incorrectly drawn) and shows just how heavily damped (unselective) the tuned circuit is under neutralized triode working, but suggests it would be better with the screened valve.[23]

A curious advertisement in Wireless World for Will Day Ltd., 19 Lisle Street, Leicester Square, London. '"The best in the West" Captain Rownd's (sic) latest screened valve as specified in the Wireless Press and acknowledged to be the most wonderful valve yet produced'.[24] Delivery in rotation was to commence on 15 September 1927.

The new valve appeared in use in the 1927 Olympia show; one Marconiphone receiver, known as the 'Round Six', used three S625 valves; Cossor had a 2 volt, 0.1 amp specimen as well as a 6 volt 0.25 amp (later advertised as 0.1 amp) example on display – both identical in appearance to the Round Form. Other manufacturers also leapt into action; and S.G. valve-types multiplied. At the Show of 1927, as well as M.O.V. and Cossor, B.T.H. apparently showed an oddity taking 0.1 amp at 1 volt on the filament, it was unique also in that it had anode and grid in a normal 4 pin base and took the screening grid contact out through the glass at the top of the valve, a curious way of re-introducing much of the anode grid capacitance, which the screening grid had removed! This specimen is mentioned in the 'Show Report'[25] but is not present in the B.T.H. advertisement in the same issue.[26] Mullard were showing 2, 4 and 5 volt types for the Manchester Show; these were all ordinary 4 pin with the anode taken out to a top-cap.[27]

It is not clear that American valve developments were much, if at all, ahead of those in Britain by now. Wireless World in June 1928 referred to the American UX222 Screened Grid valve as a '... recently ... developed ... shielded grid amplifier valve'.[28]

Secondary emission (high speed electrons knocking further electrons out of the anode which, when the S.G. voltage was higher than the anode voltage, were attracted away from the anode back to the S.G.) caused the S.G. valve to have a peculiar 'kink' in its characteristics. Conditions which could lead to instability in a circuit were easily encountered in normal working if large signals were applied to the valve. Attempts were soon made to overcome this usually undesirable feature, and so produce a form of valve suitable for output stages. The first successful specimen would appear to be by Mullard; mention of the 'pentode' (as the new valve came to be called), came in Wireless World in connection with the 'Westminster' Transportable Radio-Gramophone Receiver which used a PM 22 as a power output stage.[29] Just how such a relatively obscure firm managed to be first in the field in the use of the valve is a mystery. The new valve is discussed in an article by W. I. G. Page where it is stated that 2V and 4V types are to be produced but would not be available for some months (August/September). Reference is also made to M.O.V. whose pentodes would be available 'shortly before the Olympia Exhibition'.[30] Other

iv. ADVERTISEMENTS. THE WIRELESS WORLD AND RADIO REVIEW JUNE 6TH, 1928.

**Marconi Valve
Type H.L. 210.
General Purpose
valve.**

Fil Volts - 2.0 max.
Fil Current - 0.1 amp.
Anode Volts - 150 max.
  Amplification Factor 6.15.
    Impedance 25,000 ohms.

**Price
10/6**

THE IDEAL
CHANCELLOR

Economy. Axe-ing the amps. Getting the very last pennyweight of value out of every pennyworth of power. Keeping *your* pocket padlocked as tight as the Bank of England. Think of a Chancellor of the Exchequer like that !

That is the Marconi Type H.L.210 General Purpose valve. It uses very, very little current—only .1 amp at 2 volts.

If you will fill in the coupon below we will send you full particulars of Type H.L.210 and other Marconi Valves. Perhaps, too, you would like a copy of that handy book " 500 Marconi Valve Combinations."

MARCONI VALVES

Dept. P. The Marconiphone Company, Ltd.
210-212, Tottenham Court Rd., London, W.1
Please send particulars of Type H.L.210
and other Marconi Valve literature.

Name ....

Address ....

w.w 19

manufacturers had at least some specimens available for display.

More valve developments, of course, took place beyond this 1928 high-water mark; R.F. pentodes, multiple valves (double diode triode, double diode pentode): special frequency-changers for superhets; later T.V. types, miniaturisation; V.H.F. valves etc., but for all that the major revolution took place in 1927–28. Very rapidly the old familiar untuned triode H.F. 'amplifier' stages were done away with; gone even were the neutralized triodes, reflex amplification, mysterious detector systems, multiple R.F and L.F amplifier stages, all were discarded over a period of months. Out with them went all the familiar atmosphere in which a limited number of standard components; coils, capacitors, chokes, resistors, transformers and especially valves, could be endlessly recycled into more and more exotic and unlikely circuits, any one of which was likely to be at least as effective as anything commercially available. Instead was a brave new world full of unfamiliar components and concepts, where the rough-and-ready methods of former years could no longer play a useful part.

In earlier days home constructors and indeed manufacturers regarded valves as accessories, servants to be used, or not, at a whim; by 1929, they had become the masters, dictating all other aspects of receiver design. In fact in a Wireless World article entitled 'New Lamps for Old', the author H. B. Dent, remarks '. . . improvements in the design of . . . valves . . . will mark a new era in . . . design . . . and much of the old practice will . . . die out'. 'More often than not a wireless receiver is regarded as a conglomeration of parts,'. . . In all modern receivers the valves are the pedestals on which the design is built . . .'.[31] His article is chiefly concerned with the difficulties involved in simply trying to substitute new types of valve for old-stagers; the difficulties may be so great that it would be better to start again.

For changes in valves, then, the scope for the old-time amateur and experimenter became radically altered in a few short months in 1927–28.

References

 1 Wireless World, 6 August 1924, p. 551
 2 Wireless World, 9 September 1925, p. 318
 3 Wireless World, 26 January 1927, pp. 115ff
 4 Wireless World, 14 July 1926, p. Adv. 6
 5 Wireless World, 2 February 1927, back cover
 6 Wireless World, 26 January 1927, pp. 115ff
 7 Wireless World, 9 February 1927, back cover
 8 Wireless World, 16 March 1927, p. 330
 9 Wireless World, 9 February 1927, p. 177
10 Wireless World, 9 March 1927, p. 296
11 Wireless World, 9 April 1927, supplement
12 Wireless World, 27 April 1927, p. 526
13 Wireless World, 20 July 1927, pp. 83ff
14 Wireless World, 10th October 1928, editorial
15 Wireless World, 25 September 1929, p. 331
16 Wireless World, 27 June 1928, p. 687
17 Wireless World, 29 August 1928, p. 263

18 Wireless World, 9 November 1927, p. 656
19 Wireless World, 11 April 1928, p. 397
20 Cassell and Bernard-Jones Publications, London
21 Wireless World, 31 August 1927, pp. 260ff
22 Wireless World, 31 August 1927, pp. 267ff
23 Wireless World, 7 September 1927, p. 308
24 Wireless World, 21 September 1927, inside back cover
25 Wireless World, 28 September 1927, show report
26 Wireless World, 28 September 1927, p. Adv. 19
27 Wireless World, 2 November 1927, p. 608
28 Wireless World, 27 June 1928, p. 687
29 Wireless World, 16 May 1928, p. 535
30 Wireless World, 4 July 1928, pp. 7ff
31 Wireless World, 10 October 1928, pp. 495ff

Chapter 5
Receiver developments in America, Germany and France

The annual reports of the New York Radio Fair carried by Wireless World, usually in October and November, along with other occasional articles, allow some idea of the American radio scene in the period 1926 to 1932 to be obtained. The Annual Shows in Berlin and Paris are similarly usually reported, although not in the same detail. For the purposes of comparison with trends in Britain it is interesting to survey these reports.

During this period in America, which included the Wall Street Crash of 1929, the number of manufacturers represented at the successive shows fell from 300 to 122, similar to experience in Britain.

For 1926, the report carries the statement 'Radio has been elevated from the position of a toy to be occasionally played with . . . to the status of a real and welcome entertainer' and, curiously, similar sentiments are expressed one year later in connection with the situation in Britain, by Vincer-Minter. Another interesting comparison is that the American 1926 show had not even one crystal set on display; that situation did not arise in Britain for another year at least. These pointers, along with other slight pieces of information, suggest that America's two year lead in broadcasting was already being eroded.

In order to cope with the congested state of the American wavebands, the typical American receiver had multiple-tuned R.F. stages, perhaps 3 valves with 5 tuned circuits before the detector; mass production methods had facilitated this by providing highly consistent coils of closely matched inductance and efficient multiple 'gang' tuning capacitors so that, at most, two main tuning controls were needed. This is consistent with the image of radio as a mature source of entertainment; in such multiplicity of valves for selectivity and the organization of the receiver to make it convenient for a layman to operate, American practice was several years ahead of Britain. Other American trends were similar to those subsequently to be found in Britain; a considerable use of battery eliminators, the introduction of a few valves with thick filaments suitable for raw A.C. heating, even the use of some low consumption filament valves run in series through a rectifier and smoothing circuit from the mains – again at least a year ahead of Britain.

Freed-Eisemann eight-valve frame aerial receiver operating from the mains, American, 1927

The 1927 'New York World's Fair' report in Wireless World contains the statement 'The metamorphosis from apparatus to furniture is complete' – a feature which took another year or two to develop in Britain, and certainly never on the scale of the U.S. as is seen in the sets illustrated. There is a further statement 'The technical objectives have obviously been simplicity, selectivity, and quality'[1] which is largely an echo of what appeared in the previous year. Indeed, a survey of the sets reported upon bears this out. Only in one area was there some real progress; some 30% of the receivers used A.C. power for valve heaters, divided equally between thick filament and I.H. types of valve. In Britain for 1928, one year later, there was only about 17% mains operation; there was never any appreciable use of 'thick filament' types in commercial sets.

An article in 1928 dealing with 'American Sets of To-day'[2] gives a very similar picture with only a brief reference to the recently developed S.G. valve. This similarity of available sets over such a considerable period – 8 months, October 1927 to June 1928 – suggests some form of stagnation; the 1927 report possibly gives the reason in that it refers to the difficulty in clearing up the patent position and suggests that this had delayed progress.

The year 1929 seems to be another period of relative stagnation, in design, if not in methods. One comment was that '. . . the American broadcast receiver has almost become a standardized article . . .', it was now mass-produced and differences amongst manufacturers were small. Multiple-tuned R.F. amplifiers were still the order of the day, very heavily screened and now almost exclusively using S.G. valves. Almost invariably there was single-knob tuning, sometimes with a trimmer for the aerial circuit, perhaps only one other knob – volume control – but most sets were now mains operated. There were a few innovations; automatic volume control (A.V.C.) was tentatively being applied and moving coil speakers were more or less standard. The writer concluded that '. . . In the design of radio receiving sets the American manufacturers are, in some respects, ahead of the British'[3] and he also referred to the high level of wealth which resulted in a neglect of cheap sets. The summing up given is most probably accurate, but it does look as though British practice was catching up; in Britain at this time mains sets had climbed to some 40% of the total and 'ganged' control of the tuning was available in over 50%; the differences remaining were substantially less than one year's developments.

The 1930 report continues to show a definite lull in American progress, but economic conditions can hardly have been good. The usual multiple-tuned H.F. stages were still present, there was a move to improved A.C. 'hum' elimination, a more general application of A.V.C. (the basic principles of which were misunderstood by the Wireless World reporter), more gramophone pick-ups were available, and simpler more compact designs were evident. The changed economic conditions had shown themselves in one development; over 50 of the 176 manufacturers had introduced self-contained 'midget' sets. One straw in the wind suggesting future developments is the reappearance of the superhet; one or two had been shown with S.G. valves in I.F. stages. These sets were both more sensitive than former superhets and suffered from fewer drawbacks.

A special article 'Recent Developments in America' discusses '. . . three major developments, for all of which the Radio Corporation of America, as the virtual monopoly patent holders, may be said to be responsible:

(1) the return of the superheterodyne;

(2) the still further development and perfection of T.R.F. (tuned radio frequency) sets; and

(3) the introduction of home recording into all electric gramo-radio sets'.

The article continues, concerning the superhet '. . . the patent position was so obscure that . . . nobody was willing to risk his money on the manufacture of a set . . . the patent situation has been definitely cleared up, and the R.C.A. has the monopoly of superheterodyne patents'[4] and was, apparently, demanding very high fees for licences to manufacture. This freeing of the patents transformed the American radio scene; in the 'New York Show Report' reference is made to the fact that the vast majority of sets were already superhets with a virtual disappearance of T.R.F. It is interesting to note that similar conditions appeared in Britain in time for the 1931 September Show, some six or seven months later, although the banishment of the T.R.F. circuit in Britain was not complete – nor would it ever be. The American lead in set design, plagued by patent litigation and an unco-ordinated broadcasting system, had slipped to little more than six months. Other features of the U.S. scene were that variable mu valves (in which amplification could be smoothly controlled by the applied negative grid voltage) had come into common use and many sets had some form of visible tuning indicator, desirable in A.V.C. fitted, highly selective, sets. The economic situation had taken its toll in other ways too; cabinets were smaller and plainer, the very large pieces of furniture so common in former years were heavily marked down in price.

The impression left by this survey, and it is no more than an impression, is that the rate of progress in design in America was relatively slow over this period, particularly so from 1927 to 1930. Part of this apparent inactivity could be due to the growing economic crisis, but it seems to reflect more particularly the delays and frustrations caused by time-wasting legal arguments over patents. When the situation was finally cleared up, it coincided with a return of some economic stability and progress was extraordinarily rapid. The former universal multiple-tuned T.R.F. circuits were swept aside and replaced by more-or-less fully developed superhets in less than a year. This was a transformation of no mean magnitude; possibly much development was undertaken before licences were available – there is just a hint that this in itself had a bearing on British design and development.

The resurgence in development was short lived and soon checked by economic problems. Wireless World of 1932, in describing 'New York's Ghost Show', declared '. . . the American radio industry has gone broke'. The official show was in fact cancelled and only a small trade exhibition was held in an hotel. At this somewhat unrepresentative gathering it appeared that where sets were exhibited at all, superhets reigned supreme, although there were a few new 'straight' T.R.F. sets, redesigned round new valves. The latest fashion for midget sets was well evidenced, prices were low, but tending to rise. There was '. . . a sort of desperation in the air . . .'.[5] The effect of this new setback on the American scene appears to have been to reinforce the already obvious move to intensive mass-production. Commenting on the New York Show for 1934 Wireless World said that 'Standardisation has progressed to such an extent that one set looked very much like another, and nearly all looked like tombstones'.[6]

Progress in Germany, as reflected in the Berlin show is much more difficult to follow as reports were less detailed. Enough appears, however, to make it clear that German design, too, was strongly affected by external circumstances. In the case of Germany the financial crisis lasted much longer and was very much more severe. The German wireless industry had a smaller base to build on than its British counterpart, licence numbers in Germany were always somewhat lower and, in any case, German industry was bled to provide reparation payments in the aftermath of the War. There could be nothing of the lavish ostentation of the American radio, similarly there was little of the somewhat old-fashioned solidity of the typical British apparatus. Germany had to develop relatively simple sets using mass production methods – for which, of course, such sets were ideal – giving a supply of very cheap, efficient, if largely unadventurous, products catering for the needs of an impoverished populace. The sets were simple, easy to use, adequate for their immediate purpose but with little in reserve. Germany, too, moved toward modern superhets, but was certainly a year or more behind America and Britain. One very interesting development which significantly affected future electronics development was referred to in the Berlin Show review of 1932.[7] This was the discovery by Hans Vogt, of iron dust-cored materials which could be used as cores for R.F. coils, giving them a much higher efficiency and Q.

In Germany there was a greater political upheaval than in the other countries under review and this cast a shadow over the radio industry as well as over other aspects of life. Typical of the new order was the appearance of the German 'People's Receiver',[8&9] a two-valve set (with a corresponding three-valve battery model) built in common by twenty-eight different manufacturers, intended for reception of the nearest (German) high power station and 'Deutschlandsender'. Hitler's propaganda machine was in action! This determination to be heard had some effect on transmitter policy which contributed to the chaos and was only partially solved by the various international agreements.

France appeared to be very different; in contrast with Britain or America there did not seem to be all that much enthusiasm for wireless. The industry, too, was protected from outside influences long after Britain abandoned the 'BBC' mark with all the restrictions inherent in it. In fact it was not until 1930 that foreign influences could be discerned in French design. Prior to that, French manufacturers seemed wedded to the outmoded old superhet design with all its many drawbacks; designs dating

Professor Leithauser, designer of the German 'People's Receiver', with the A.C. model of his creation, 1933

back to the early 1920s were common. Reasons given for this state of affairs do not seem to be all that convincing; it was apparently difficult to obtain permission in Paris to erect an outside aerial; hence the need for a sensitive frame-aerial receiver. French listeners also seemed to have a predilection for foreign broadcasts, tending to ignore what there was in the way of local programmes; again a situation favouring one of the old style superhets. A general inertia in the industry, a lack of desire to change, may however, have been a considerable factor. There were signs of change by about 1929. A year later American sets began to enter the country (requiring long wave facilities to be added), and even some signs of British influence could be seen. The ancient superhet was very rapidly rivalled by multiple-tuned R.F. stages on the American pattern.

French design, then, seems to have been constrained by a poor home-market, deliberate isolation and a general lack of enthusiasm; it was several years behind America and Britain in 1930; but by 1933, the French were well away in pursuit of American practice. The Paris Show was reported to have a predominance of 'Gulliver' sets; midgets, mainly of superhet circuitry, but in a modern form. It seems unlikely that the French radio industry affected British design in any great way, although there was a brief flirtation with the French 'bigrille' two grid valve, almost universally used in France for frequency changing in superhets. But there was a considerable amount of evidence indicating that French broadcasting influenced British policy to a large extent. At a time when the typical B.B.C. fare on Sundays consisted mainly of church services and talks which did not clash with church times, in France, Fécamp, (Radio Normandie), ran a continuous service of (sponsored) dance music and other forms of entertainment, a regular thorn in the side of the Reithian B.B.C.

It appears from this survey that in the U.S.A. and on the Continent as well as in Britain, the culmination of many years of slow evolution was the rapid, almost universal, adoption of the modern superhet, and that the revolution took place in the few years around 1929–30–31. The whole wireless world fell under the spell of the superhet; a mention in the Wireless World report of the Brussels show contains a phrase which seems to sum up the situation '. . . a triumph for the superhet'.[10]

References

 1 Wireless World, 12 October 1927, p. 513
 2 Wireless World, 27 June 1928, pp. 685ff
 3 Wireless World, 30 October 1929, pp. 485ff
 4 Wireless World, 11 February 1931, pp. 138ff
 5 Wireless World, 28 October 1932, p. 402
 6 Wireless World, 26 October 1934, p. 330
 7 Wireless World, 16 September 1932, p. 281
 8 Wireless World, 11 August 1933, p. 111
 9 Wireless World, 8 September 1933, pp. 215ff
10 Wireless World, 15 September 1933, p. 233

Chapter 6
British domestic wireless, 1924–34

During this decade, British wireless underwent a complete transformation; it is the purpose of this chapter to consider some of the influences bearing on that transformation.

As we have seen, so far as the purchaser of equipment was concerned, the situation in 1924 had not changed in any essential way since the inception of broadcasting. We have already looked at the changes in transmitting conditions over the following decade and briefly at techno-logical change; but we must remember that this was also an era of great social, economic and political change; the time of the 'slump', the General Strike and the 'great depression'. These events, also, affected the development of wireless throughout this short period; some historical background is necessary if we are to attempt to understand the sequence of events.

One reliable method of examining the growth of the wireless industry is available to us; the number of wireless receiving licences in operation at any given time. A graph is appended showing this chart of progress over these ten years. It is clear that this graph shows a largely unbroken climb from about 600,000 in January 1924 to 6,600,000 by December 1934; it is also clear that it is by no means a uniform growth. Three phases are quite distinct:

(1) 1924 to the second half of 1926: licences increased at an average rate of 670,000/year.

(2) End of 1926 to second half of 1930: licences increased much more slowly, an average of 225,000/year.

(3) End of 1930 up to 1934: very rapid growth, licences increased at an average rate of 893,000/year.

Within these broad movements small deviations may be noticed; it is probable that these reflect a fall-off of interest in the summer and an awakening as winter approaches in an annual cycle.

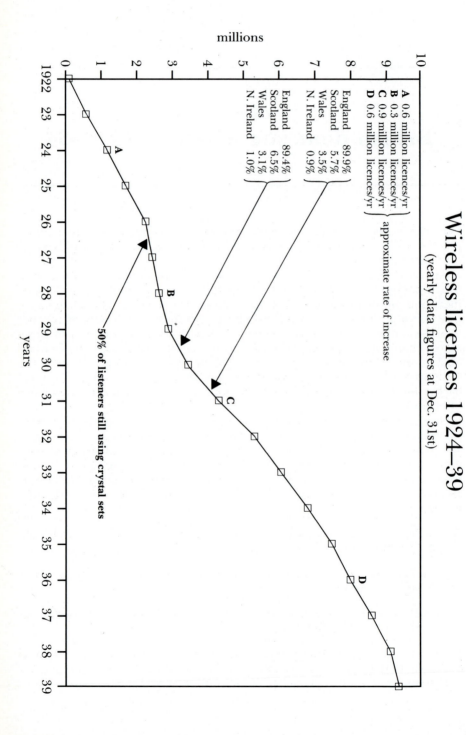

Wireless licences 1924–39

(yearly data figures at Dec. 31st)

A 0.6 million licences/yr
B 0.3 million licences/yr
C 0.9 million licences/yr
D 0.6 million licences/yr

> approximate rate of increase

England	89.9%
Scotland	5.7%
Wales	3.5%
N. Ireland	0.9%

England	89.4%
Scotland	6.5%
Wales	3.1%
N. Ireland	1.0%

50% of listeners still using crystal sets

millions

years

To explain the large-scale changes, it is tempting to think in terms of the general economic activity of the country, times of slump and recovery, but this is a facile assumption which requires examining in the light of the history of the period.

Historical background

The political and economic history of the U.K. at this time, played out in the aftermath of the Great War, also shows several distinct phases. The first is the immediate post-war period up to 1922, characterized by optimism gradually deflating into disillusion and ending in the economic 'slump' of 1921 when the unemployment rate reached the dangerously high level of 22.4% (July 1921). The Coalition Government of Liberals and Conservatives fell in 1922; in the next three years there were three administrations.

The first of these was Conservative, originally under Bonar Law, later under Stanley Baldwin; it failed to carry a Protective Tariff Bill and resigned. The subsequent general election in 1923 produced an uneasy alliance of Labour and Liberal, with Ramsay MacDonald as Prime Minister. This endured for only 10 months and made little impact on the economic situation of the country. Political stability was not restored until the election of October 1924 when the Conservatives under Baldwin triumphed; this administration was in a powerful position and lasted out its full term until 1929.

In April and May 1926, a steadily deteriorating industrial scene flared up into a major confrontation between the Government and, initially, the miners. This soon involved the whole Trade Union movement and led to the General Strike 'The predominating event of the year'.[1] There was much talk of 'Communist Revolutionaries' but with the exception of the mining industry, little evidence of any desire for revolution showed amongst the strikers. In the mines, however, communist influences were at work; money, advice and help were rumoured to have found their way to the mining unions from Russia. The General Strike was short-lived; many factors contributed to its downfall amongst which might be mentioned the fact that most of the strikers had seen the desirability of discipline during war-service and the Government's appeal for loyalty amongst non-strikers resulted in a massive volunteer work-force being rapidly mobilised. Food and fuel supplies were ensured by strong Government action, even some mail services were maintained; quite rapidly the T.U.C. was forced into a humiliating climb down. One major factor in this government victory was the use it was able to make of the B.B.C. in the absence of virtually all daily press except for Winston Churchill's British Gazette! 'The British Broadcasting Company issued regular messages of offical and other news'. 'The sense of detachment was overcome by the bulletins of the B.B.C. which ... were in most places multigraphed or copied for display on shop windows, etc.'. On Saturday 8 May 'Mr Baldwin broadcast a message to the nation – the first occasion

IV.   ADVERTISEMENTS   THE WIRELESS WORLD AND RADIO REVIEW   MAY 26TH, 1926.

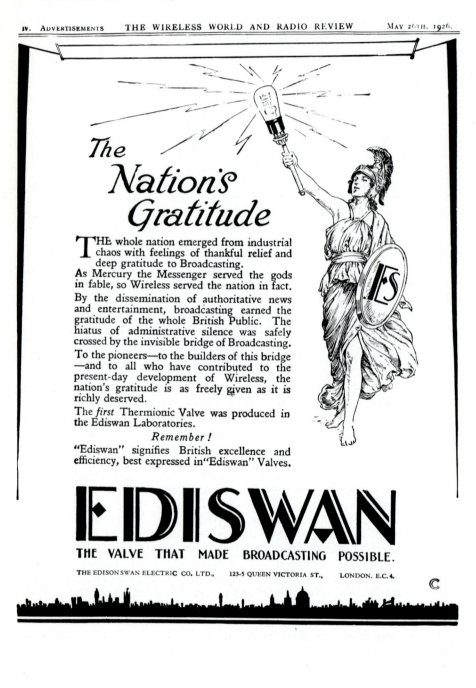

when a Prime Minister thus used wireless communication in a national crisis'. Later, 12 May 'At night a long message from the Prime Minister was broadcast and joyful celebration of the end of the strike was general'.[2] Wireless had clearly come of age and shown itself to be a most significant instrument of state. The lessons learned in this conflict are neatly encapsulated in an advertisement by Ediswan headed – 'The Nation's Gratitude'. 'The whole nation emerged from industrial chaos with feelings of thankful relief and deep gratitude to Broadcasting. . . . The hiatus of administrative silence was safely crossed by the invisible bridge of Broadcasting . . .'.[3] Just how far-reaching were the consequences of this first successful use of broadcasting in a crisis is difficult to judge; the then Chancellor of the Exchequer, Winston Churchill, certainly appears to have remembered the episode; he was himself to make remarkable use of the power of broadcasting some 13 years later. Asa Briggs covers the topic,[4] and there are subsequent events which may indicate that the significance had not escaped official thinking.

The first of these was in the wording, or possibly re-wording, of a passage in the new B.B.C. Charter, effective from 1 January 1927, as reported in Wireless World '. . . a provision had been inserted in the Charter that the Corporation should broadcast matter if requested to do so by a government department. An important provision was that of the resumption (sic) of full control over the service by the Government in case of a national emergency – an absolutely necessary and vital provision'.[5]

The other episode is rather odd. P. P. Eckersley had for years been trying to get the Regional Broadcasting scheme under way; we have seen how the original 5XX was authorized, experimentally, as far back as 1924, but there had never been authority given for any further advance. Sometime in the summer of 1926, however, permission was granted for the second phase of the scheme (the 5GB transmitter), again on an experimental basis. This decision seems to have been shrouded in secrecy, however, for it was not until 12 January 1927 that even Wireless World carried a first report[6] and, incorrectly at that, by which time the transmitter was within a few weeks of being operational.

At the conclusion of the strike, workers in some industries (coal, railways), were forced to accept wage-cuts as part of the return-to-work arrangements. In spite of this and the general economic consequences of the strike, the three years 1926–29 were a relatively tranquil and reasonably prosperous interlude. Unemployment had halved, levelling out at 10%, and production had slowly risen to above the 1913 level. The Government, however, was passive and failed to tackle the still underlying economic problems; in the 1929 election it lost much support to Labour and even the Liberals managed a revival. Ramsay MacDonald once again formed a mild Labour administration, but the new Government was not helped by the Wall Street Crash and its efforts did not stem the tide of mounting unemployment – 23% by mid 1931. So severe was the drain on the Unemployment Insurance Fund that a budgetary and financial crisis even threatened the country's international credit.

The situation deteriorated further and MacDonald resigned in August 1931; however, neither Baldwin nor Samuel (the Liberal leader) was prepared to form a government and eventually MacDonald was persuaded to form a Coalition, the 'National Government', which took office on 24 August 1931. Having secured the essential international credit the Government called fresh elections. The National Government, fighting under the slogan 'Safety and the Union Jack', was confirmed in office by a massive majority; the Labour Party, which had repudiated Ramsay MacDonald, fought independently but suffered a crushing defeat. The new administration pushed through a Bill to give a General Protective Tariff and, although the Depression continued even beyond 1933, conditions quite rapidly improved. The drop in British production from 1929 to 1932 was 16% but by 1937 it had reached 50% above the 1932 level; British industry turned largely to its home market as the recovery was not seen in overseas trade which continued to decline.

To summarize, the very worst periods for the economy were (a) starting in 1921 and festering on until 1926, (b) 1930 and a year or two after: yet these are the very periods when licence figures rose most rapidly. Furthermore, during the interlude from 1926 to 1929, the period of relative stability and prosperity, however illusory it might have proved to be, the licence figures increased only very slowly. The large scale trends in the licence figures appear to be entirely opposite to what might be expected.

As a coda to this section, it is useful to consider just what sort of a market was available to wireless manufacturers at some appropriate points along the graph. In any one of the years 1927 to 1929, the increase in the number of licences was of the order of 225,000 and this represents the potential number of new customers for apparatus. To this must be added existing licence holders changing their sets. As we shall see later, however, there was not a great incentive to change at this period as really new developments in manufactured equipment were not forthcoming, so this is likely to be a small number at best. The other influence is the unknown number of home constructors at work. If the evidence of the number of wireless societies in existence is anything to go by, this must be a substantial number. Taking these two factors into account, a figure of 200,000 is likely to be a fair representation of the potential sales for complete sets for the whole wireless industry. In the Wireless World 'Buyer's Guide'[7] for 1926 some 100 manufacturers are given with a total of well over 500 sets on sale; thus the 'average' manufacturer could look for sales of only 2000 sets and production runs of under 1000 on any one model. Many manufacturers at this time were of only local significance, with production runs possibly numbering dozens rather than thousands.

Moving on to 1934, the corresponding figure for potential sales was over 800,000, but there were now just 43 manufacturers listed,[8] and although there appear to be 226 different sets available, close inspection reveals that many of these were either the same chassis mounted in a different cabinet, e.g. a radiogram rather than a table model, or a console; in some cases very similar chassis were used for A.C. only or A.C./D.C. models. There were also a number of very expensive receivers – over 100 guineas – whose sales must have been very small. Allowing for all these different factors, there would seem to be no more than about 80 genuinely different sets for the mass market. The 'typical' manufacturer could by now expect to sell some 20,000 sets in total with production runs approaching 10,000 of any one chassis.

These figures give some indication of the transformation in the British wireless scene in the decade to which reference was made at the start of this chapter. We shall now examine the progress of the wireless manufacturing industry in a more detailed way over the period.

The trend in receivers

There is no doubt that, in 1924, by far the majority of listeners used crystal sets with headphones and had a substantial outside aerial; the whole B.B.C. policy was directed at providing a service in populous areas suitable for crystal reception. (Eckersley's 5mV/m field strength contours). There was no shortage of commercially available crystal sets; literally hundreds of manufacturers were in production, many being small concerns which disappeared rapidly from the scene, but the large, household-name manufacturers were all represented. Prices varied widely; a comprehensive receiver such as the 'A.G.F.' of 'effective range 50 miles', mounted in a mahogany box and complete with headphones, aerial equipment etc., retailed for £4.12.6; a simple tubular coil with cat's whisker detector could cost as little as 10/6, but to this would have to be added headphones, aerial and earth, insulators, wire, also B.B.C. Royalty payments (which ended in 1924); all complete, the cost would be about £3. Valve sets, too, were expensive, and were mostly advertised without rather essential 'accessories' such as the valves, headphones, batteriees, aerial and earth equipment, Marconi Royalties (12/6 per valve holder, in addition to B.B.C. Royalty payments which ended in 1924); thus giving a somewhat misleading picture of the expense. The habit of advertising prices exclusive of accessories follows naturally when it is realised that the manufacturers usually did not produce the component parts themselves, but relied upon outside suppliers for these items which they then assembled. As we have already noted, there was not a great deal of difference from one valve to another at this date; most sets would work with most valves, from most manufacturers. Similarly with head-phones and loudspeakers; there were very large numbers of different models produced by specialist makers and it was largely a matter of personal preference whether to use several pairs of headphones for a family or to listen to a loudspeaker. A certain economy was possible, too, in that many listeners would have a set of headphones or an odd valve or two to hand from previous wireless sets. When all essentials are added in, costs for a large number of receivers advertised in Wireless World for 1923–24 average out to: 1 valve £12. 2 valve £18. 3 valve £25. 4 valve £29.

As we have seen, no very significant changes occurred in broadcasting between 1924 and 1926, other than the introduction of 5XX. In receiver design, apart from some valve developments and price reductions, sets were recognizably similar, although neutralized triode H.F. stages were becoming common. It is much easier to glean a complete picture of the domestic wireless scene for 1926 since Wireless World published a 'Buyer's Guide'[9] for receivers currently on the market. The richness of the scene may be judged from the bare statistics; for crystal sets some 51 manufacturers are listed and 109 sets described. The 'average' price for a complete, operating, crystal set was now somewhat less than £3, so prices were tending downwards. The number and variety of crystal sets may appear surprising, but it has to be remembered that even as late as mid 1927, 50% of all listeners used them. There were, in addition, a substantial number (24) of crystal/valve combinations. Several of these employed more or less complex reflex circuitry in which still-expensive valves, rendered even more costly by the Royalty, were made to do double duty and amplify at both radio and audio frequencies. A number of others were no more than two separate units, a crystal set coupled to a one or two valve 'note magnifier', a practice which had been very common in the very earliest days of Broadcasting. It allowed a start to be made with a relatively inexpensive crystal set, to which more amplifiers could be added as finance and inclination required.

In pure valve receivers, some 64 manufacturers are named and no less than 470 receivers. When allowance is made for all essential 'accessories' the following figures for 1926 emerge in Table 2:

Table 2

Valves	Number	%	Average cost
1V	41	9%	£8
2V	139	30%	£14
3V	126	27%	£23
4V	128	27%	£38
Over 4V	36	7%	Wide Ranging

These figures are interesting in that 2, 3 and 4 valve sets are all equally favoured and apparently quite dominate the market.

Compared with the corresponding figures for 1924, we find quite substantial falls in the prices of 1 and 2 valve sets, perhaps a slight reduction for 3 valves and a substantial increase for 4 valves. The drop in price of 1 and 2 valve sets may just reflect a general reduction in the cost of components, valves especially, but it was quite a considerable reduction and was greater than it appears as many of these sets were fitted with at least an extra coil holder for allowing them to 'tune up' to the long wave, 5XX transmissions. It was also noteworthy that, for a 2 valve set, the Marconi Royalty was becoming a significant proportion (10%) of the price. The large increase in price for 4 valve sets was most probably due to the fact that, in this price range, they were a real luxury item, probably only to be bought by the wealthy, for whom a higher degree of finish, a more elaborate cabinet in a fine wood, possibly even a cabinet with built-in loudspeaker, would be an attraction and the extra cost of much less relevance. There was also a considerable selection of 'amplifiers', 55 are listed; as mentioned above, these were intended as 'add on' units to allow an existing crystal or single-valve receiver to operate a loudspeaker. Thus the significant trends between 1924 and 1926 include the following:

Manufacturers began to hide more of the valves, coils and condensers within a more or less elaborate cabinet. Except for portable receivers which were becoming common in 1926, very little effort had been made to locate loudspeakers and batteries within the container. There were also the beginnings of an attempt to produce sets which could be operated partially or entirely from the mains, although no special valves for this purpose were available and most of the schemes used were, to some extent, makeshift.

The two years 1927–28 also did not produce great changes in set design; but the new valves which became available, (indirectly heated KL1, January 1927, screen grid S625 September 1927), laid the foundations for the rapid changes which were to follow.

The first sentence in the 1927 Wireless World article 'The Trend of Development' is 'Many real advances with steady progress everywhere'.[10] The 'Guide to the Show' is more specific 'The progress made on the transmitting side is largely instrumental in setting the pace for receiver development . . .'[11] and this theme is enlarged upon (in the first mentioned article) where the writer maintains that '. . . the greatest improvement has been made in the matter of selectivity. . . . when the Regional Scheme comes into operation a set which now gives satisfaction . . . will become hopelessly out of date, . . .'[10] (5GB came into action on 21 August 1927 on 30kW). The screen grid valve, which had the potential to deal with the selectivity problem, had arrived on the scene too late in the year to exert its future overwhelming influence on design, although the Marconiphone 'Round Six' was on display and claimed to be 'An outstanding example'. The set itself, however, seemed to be distinctly unambitious; it contained 3 S625 valves, but the stage gain of each was apparently little, if at all, greater than that of a standard neutralized triode. In addition, the amount of screening apparently necessary must have added greatly to the complexity and cost of manufacture. Furthermore, the four separate tuning controls, even although mounted edgewise in pairs, must have made station-finding a distinctly daunting task. Other sets using the S625 are mentioned; the B.R.C. 'Radio Exchange' (incorrectly described as the 'B.B.C'!) and Wireless World's own 'New Everyman Four',[12] was apparently available commercially from Peto-Scott, who normally specialized in producing kits of parts, to an author's specification, ready for home constructors.

iv. ADVERTISEMENTS. THE WIRELESS WORLD AND RADIO REVIEW FEBRUARY 9TH, 1927.

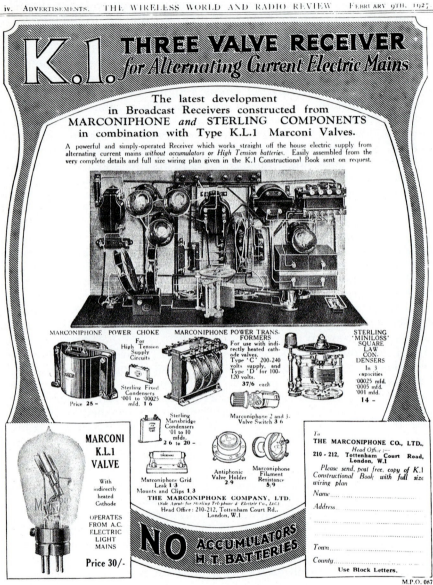

K.1. THREE VALVE RECEIVER
for Alternating Current Electric Mains

The latest development in Broadcast Receivers constructed from MARCONIPHONE *and* STERLING COMPONENTS in combination with Type K.L.1 Marconi Valves.

A powerful and simply-operated Receiver which works straight off the house electric supply from alternating current mains *without accumulators or High Tension batteries*. Easily assembled from the very complete details and full size wiring plan given in the K.1 Constructional Book sent on request.

MARCONIPHONE POWER CHOKE
For High Tension Supply Circuits

Price **25 -**

Sterling Fixed Condensers ·001 to ·00025 mfd. **1 6**

MARCONIPHONE POWER TRANS-FORMERS
For use with indirectly heated cathode valves.
Type 'C' 200-240 volts supply, and Type 'D' for 100-120 volts.
37/6 each.

Sterling Mansbridge Condensers ·01 to 10 mfds.
2 6 to **20 -**

Marconiphone 2 and 3. Valve Switch **3 6**

STERLING 'MINILOSS' SQUARE LAW CON-DENSERS
In 3 capacities
·00025 mfd.
·0005 mfd.
·001 mfd.
14 -

MARCONI K.L.1 VALVE

With indirectly heated Cathode

OPERATES FROM A.C. ELECTRIC LIGHT MAINS

Price **30/-**

Marconiphone Grid Leak **1 3**
Mounts and Clips **1 3**

Antiphonic Valve Holder **2 9**

Marconiphone Filament Resistance **5/9**

THE MARCONIPHONE COMPANY, LTD.
(Sole Agents for Sterling Telephone & Electric Co., Ltd.)
Head Office: 210-212, Tottenham Court Rd., London, W.1

NO ACCUMULATORS H.T. BATTERIES

To
THE MARCONIPHONE CO., LTD.,
Head Office :—
210 - 212, Tottenham Court Road, London, W.1

Please send, post free, copy of K.1 Constructional Book with full size wiring plan

Name..............

Address..............

....

....

Town..............

County..............

Use Block Letters.

M.P.O. 087

Regarding indirectly heated valves, Wireless World comments 'Valves with indirectly heated cathodes . . . are gaining in popularity, . . . It would seem, however, that these valves are only of interest to the home constructor, and that they have not yet found their way into commercial receiver design.'.[13] Strange words, indeed, for the KL1 was advertised in February 1927 and the Marconiphone Company was offering a complete kit set, their 'K1 Three Valve Receiver' said to be 'Easily assembled . . .', in an advertisement.[14] The much more satisfactory Cosmos series only put in an appearance at the Show, so could not have been incorporated in new designs by that time. It is possible that both the Osram innovations were a little ahead of their time; it is even more possible that they had been hurried onto the market before their development was far enough advanced. Certainly there are hints of unsatisfactory variations in characteristics and of other problems in the months to come.

Other than these, possibly unwise, innovations, there is not a lot to comment on in the receivers exhibited in 1927. One or two telling phrases emerge from a careful study of the 'Trend of Development' article: '"Ganging" of condensers, which gives a single-knob tuning control, is increasingly popular, . . .' and is thought to be likely to become standard '. . . except perhaps for those who are willing to learn enough about their sets to enable them to operate the more complicated arrangement.' '. . . practically every maker has decided that more than one tuning control is beyond the ability of the average non-technical purchaser. . .' '. . . the cone or diaphragm loud-speaker has firmly established its predominance over the once popular horn type, . . .'. 'The one-time ubiquitous plug-in coil shows signs this year of waning popularity . . .'. 'Battery eliminators. . . . Not only has experience led to greater confidence in the success of the eliminator, but the requirements of the modern set . . . have forced the adoption . . .'.[15]

The impression derived from this article does not differ very markedly from that portrayed for America at this time except perhaps that in America 30% of all receivers were already fitted for all-mains operation, more appropriate for a country where electricity supplies were more widespread. American receivers were also, with few exceptions, very much larger and luxurious, although prices were not correspondingly higher; the use of reaction, so typical of British practice, was almost totally foreign to American use, not surprisingly, with multiple-tuned H.F. stages and a predilection for local station listening, it simply was not necessary.[16&17]

In connection with the emergence of mains-operated receivers it is interesting to note that in 1926 the Wireless World 'Show Review' article makes the statement 'The first set to be produced in this country in which filament heating as well as anode current supply is derived from alternating current mains is shown by Gambrell Brothers Ltd.'.[18] Valve filaments, of the 60 mA variety, were connected in series across the HT supply.

Moving on to 1928 a number of trends become evident, especially when data for 1927 and 28 are studied together. The main sources for this comparison are two articles in the same edition of Wireless World 'Sets of

DECEMBER 1ST. 1926.   THE WIRELESS WORLD   ADVERTISEMENTS.   29

"THE WIRELESS SET OF THE FUTURE"

Cabinet Three Mod I.

Models and Prices.

For direct current lighting supply.

BABY GRAND, 2-VALVE SET,
£15.15.0

CABINET THREE, 3-VALVE SET,
£23.10.0.

For alternating current lighting supply.

BABY GRAND, 2-VALVE SET,
£21.10.0.

CABINET THREE, 3-VALVE SET,
£28.10.0.

A PRESENT WITH A HAPPY FUTURE.

The following is a letter received from
CHARLES DUNCAN, Esq., J.P., M.P.

House of Commons.

DEAR SIRS,

I thought you might like to know how the Wireless Set you supplied me with is functioning.

It gives me the greatest possible pleasure to inform you that it is giving me the completest satisfaction.

Its service is instant, regular, unbroken and delightful to a degree. It certainly is a revelation and leaves the **old type with accumulator and high tension battery miles out of date.**

It certainly seems to me to be without question **the Wireless Set of the future.**

Wishing you every success. Believe me,

Yours very sincerely,

(signed)   CHARLES DUNCAN.

NO ACCUMULATORS OR BATTERIES REQUIRED WITH GAMBRELL MAINS RECEIVERS.

The Baby Grand 2-Valve Set gives excellent loud-speaker reception of Local Stations and Daventry.

The Cabinet Three, 3-Valve Set, besides giving splendid loud-speaker reception of Local Stations and Daventry, will give you several alternative programmes.

Note.—The prices quoted include all valves and coils for Local and Daventry Stations. (*Marconi Royalty extra.*) Cost of running averages 2d. per week for 3 hours' use each day.

COMPONENTS THAT COUNT.

We have the courage of our convictions.

The following components which we manufacture are used in all our receivers:

L.F. TRANSFORMERS.
STANDARD COILS.
TAPPED COILS.
NEUTROVERNIA
CONDENSERS.

Illustrated Folders and full details of all the above, free from :—

The Gambrell L.F. Transformers. Give purity of reproduction and good volume.   Ratios $3\frac{1}{2}$ and 5 to 1.
Price 25/6 each.

The Gambrell " Efficiency " Coils. These standard coils are acknowledged as the most efficient obtainable.

The Gambrell Centre Tapped Coils. Fulfil every requirement of the most advanced practice in radio reception.

The Gambrell Neutrovernia Condenser. Has more good points in design, construction and efficiency than any other can claim.

GAMBRELL BROTHERS LTD., 76, VICTORIA STREET, LONDON, S.W.1.

Parr's Ad.

the Season'[19] and 'Buyer's Guide'.[20] 107 manufacturers are given \and a total of just under 400 sets are listed. The tables distinguish between 'Cabinet-type' (229) and 'Portables and Self-Contained' sets (156). It is not always easy to distinguish between the two categories in the second list; (some self-styled 'portables' would turn the scales at half-a-hundredweight!). The clear divisions into types as was done for 1926 must be viewed in the light of this observation, but the conclusions are still of interest.

It is notable that no crystal sets are listed, although some were still available. Eliminating cases where a precise allocation to one group or another is unclear, we are left with 175 ordinary receivers, 122 portables and 59 mains sets, giving a total of 356 receivers.

There were also available 14 superhets, all of them old-fashioned (4% of the total), together with short-wave receivers and a few specialist radiograms. The breakdown in numbers, percentages and prices as given in Table 3.

Table 3

1928	Portable			Ordinary			Mains			Totals		
Valves	No.	%	Price	No.	%	Price	No.	%	Price	No.	%	Price
1V	1	1	£ 5	2	1	£ 3	–	–	–	3	1	
2V	7	6	£10	61	35	£ 9	12	20	£19	80	22	
3V	28	23	£19	77	44	£17	27	46	£26	132	37	
4V	17	14	£25	19	11	£26	13	22	£33	49	14	
5V	65	53	£25	5	3	£29	7	12	£50	77	22	
Over 5V	4	3	–	11	6	–	–	–	–	15	4	
Totals	122	34		175	49		59	17		356		

Figures taken from the analysis of set design 'Sets of the Season' make interesting comparison with those in Table 3.

Table 4

Valves	1927 %	1928 %
1V	3	1
2V	21	20
3V	30	39
4V	20	14
5V	18	21
Over 5V	8	5

The 1928 figures in Table 4 should be very similar to the last percentage column in Table 3. Discrepancies are small and do reflect slight differences in assigning sets to various categories; for instance, does a BTH 2-stage valve count as one valve or two? Also, many 'mains' sets were really battery sets with eliminators allowing them to be run from D.C. mains; occasionally they had A.C. eliminators along with 'floating' accumulators or trickle chargers of some sort. The agreement, however, is satisfactory.

MAY 16th, 1928.

537

Sound Constructions and Exceptional Long=wave Efficiency.

THE Ormond is an excellent example of the cabinet-type " transportable." It is neat and compact, and its weight compares very favourably with many alleged " portables." Coming from the Ormond Engineering Works, it is only to be expected that the construction is solid and the workmanship sound.

The controls could not be simpler. A two-way switch, with central " off " position, gives high or low wavelengths, and there are two slow-motion dials for tuning and reaction.

On switching on the set for the first time one is impressed by its general liveliness ; carrier waves can be picked up by the dozen, and most of them can be resolved into telephony. The efficiency on long waves is extraordinary, and Hilversum, Warsaw, Zeesen, and Radio Paris, as well as many other long-distance stations can be received in daylight with certainty. There is a tendency to threshold howling on some wavelengths, but over the greater portion of the dial reaction control is satisfactory. The selectivity on short waves is well above the average, and two or three foreign stations can be tuned in clear of 2LO when only 1¾ miles from that station.

The ball-bearing turntable is a great boon when working near to a powerful transmitter, as the position of minimum pick-up can be accurately found with a marked increase in the apparent selectivity, and when conditions are favourable for distant reception a great improvement in signal strength is obtained by setting the frame exactly in line with the station being received.

Quality of reproduction is good, but the loud - speaker movement is non-adjustable, and will not take the full output of which the set is capable without jarring. This condition is not reached, however, before an undistorted sound level adequate to most needs is reached.

The circuit is developed along conventional lines, the two H.F. valves being coupled by aperiodic chokes covering both long and short wavelengths. The valves are mounted on a rubber-sprung sub-panel ; there was some tendency to microphonic howling when the set was first tried out, but this subsequently cured itself. Ormond components are used throughout, including the loud-speaker.

This is a receiver which immediately attracts attention by reason of its neatness, sound construction, and lively performance. When the question of price is taken into consideration the Ormond takes its place with the select two or three which remain after a process of elimination, and from which the last difficult choice must be made.

A completely detachable back secured with lock and key gives easy access to valves and batteries in the Ormond self–contained receiver.

A 49

ORMOND 5-VALVE PORTABLE.

Type : *Cabinet transportable.*

Circuit : *2-v-2.*

L.T. Capacity : *30 amp. hrs. Consumption : 0.55 amp.*

H.T. Capacity : *" Single." Consumption · 6 mA.*

Weight : *32 lbs.*

Dimensions : *18in. × 14in. × 9in.*

Price : *£27 12s. 6d.*

Maker : *Ormond Engineering Co. Ltd., 199/205, Pentonville Road, Kings Cross, London, N.1.*

In the 1928 figures the principal difference between ordinary and portable sets lies in the predominance of 5V sets in portables; almost all of these, however, are old designs using untuned H.F. triode amplifiers and not new receivers. The number of portables may appear surprising, but this probably reflects the facts that portables were usually totally self contained, easy to operate, well finished and unlikely to dominate a room. Thus the portable was the only type of British set which matched the current trends in listeners" requirements as shown by American experience.

That this was a transition period is reflected in the marked surge in numbers of 3 valve sets from 1927 to 1928; two new valves were largely responsible for this. Firstly the fully developed S.G. valve in a suitable H.F. circuit could give a total gain very similar to that possible using two neutralized triodes. Secondly, the pentode output valve (only introduced at the beginning of 1928) could be fully driven by transformer coupling from a standard 'leaky grid' detector; a triode output valve required an extra L.F. stage.

Comparing 1928 with 1926 and 1924, the steady decline in prices is most obvious and it must be realized that the situation is even more marked than it appears; a typical 1928 3 valve (S.G. det. pentode) fitted with reaction was fully able to supply all reasonable needs and it was a much better receiver than even 4 or 5 valve sets from previous years. Effectively 1928 prices were about half those of 1926, performance was very much improved and running costs were appreciably better. As mentioned above, most of the 'mains' receivers were some sort of hybrids, typical of a transition period. This era of rapid change is also highlighted by an examination of the number of sets using S.G. H.F. amplifiers; 53% of ordinary sets were already using S.G. valves; in portables it was only 27%, presumably reflecting the retention of many obsolescent designs by small firms.

It is clear that the same trends are apparent in both American and British practice, with the notable exception of numbers of valves; in Britain, the consistent use of reaction increased both sensitivity and selectivity adequately without moving to large numbers of T.R.F. stages.

Transition

From the start of broadcasting until the end of 1928, therefore, the picture is one of slow evolution. Events hailed as 'revolutionary' in various advertisements over these six years did not, in the short term, disturb the even pattern of development. But wholesale change was in the air. A quotation from Wireless World sets the scene for the future and gives a vivid picture of manufacturing procedure as it operated at least to the end of 1928 '. . . The typical British wireless receiver is an aggregate of highly finished units, any one of which could be boxed in an attractive carton and sold as a separate component. For instance, intervalve transformers complete with nickelled terminals and polished case, . . . are to be found . . . out of sight beneath panels. Apart from the extra cost, the space occupied by this redundant material frequently enforces a layout which is inimical to the attainment of maximum efficiency ... and the finished receiver is so large and conspicuous that it must be housed in an expensive piece of furniture before it can be installed in a drawing room. A breakaway from ... building sets ... with an eye on the component market has already been made, notably in America, Holland and Germany'. Continuing to discuss the particular G.E.C. receiver (featured in the 1928 Radio Show), which had sparked off the discussion, the article suggests ... 'The "Victor 3" may do for wireless reception what the Morris-Cowley did for motoring'.[21] The writer of the article clearly sees in this one receiver the seeds of the future. And yet there is a mystery. The whole description of the design and manufacture of this British set, as well as its appearance as illustrated in the article, very strongly suggests it to have been identical with a Telefunken set shown at the Berlin show of 1927, one full year earlier.[22] Certainly the German influence must have been very strong to have produced such a replica, quite unlike the other G.E.C. products typical of

Telefunken three-valve receiver, 1927

the time. However that may be, the abandonment of the 'British Only' policy inherent in the 'B.B.C' mark was, perhaps belatedly, laying British manufacturing open to influences, if not direct competition, from overseas.

Transformation

The major transformation of the British wireless industry took place, as we shall now see, during the second half of the decade under review. Big firms, the Marconiphone and G.E.C. companies amongst others, were very active, but in the main the norm, up till 1929, was for a large number of small firms constructing standard laboratory apparatus, often with valves, coils, batteries, wires, horns and more all on proud display; after this date the scene quite rapidly changed, giving way to large scale, often integrated, plants, capable of providing the first class all-enclosed receivers demanded by the public, using mass-production methods to attain more realistic prices. It is not surprising that such changes had to come first from the bigger firms and it is of interest to see what external pressures there were on the usually very conservative British manufacturers.

As well as overseas influences, two other causal forces are worthy of note; one, which we have already examined, was in valve design, where Britain was very far advanced, the other was in the changes in transmitting practice which we have also touched upon but now require to elaborate. By August 1927 Eckersley's Regional Scheme was at last well under way with Daventry 5XX in full operation on long wave, 5GB on medium wave, the Brookmans Park twins soon to follow. The numbers and power of Continental stations were multiplying; the Geneva Plan, the Brussels Plan, the Prague Plan (30 June 1929) had all attempted to clear the air; but a simple crystal set, or even a one or two valve set, could no longer select out of the mass of transmitters the one desired; such had never before been required of it. It is important at this stage to see what conditions were like overseas, on the Continent and in the U.S.A., to be able to assess what other influences were possibly at work on British design. Bad as the situation was in Europe, it had already become very much worse in the U.S.A. where broadcasting had started two years earlier, but, and much more importantly, the unbridled commercialism of American radio had ensured the maximum number of transmitters all shouting their loudest. Such pandemonium had early required special measures to ensure adequate selectivity. The absence of the Marconi Royalty had its influence, too, as it allowed American manufacturers to be profligate in the use of valves. As is stated in a separate chapter, American practice tended to be towards large, multi-valve sets, with numerous H.F. tuned stages for selectivity. This combination of circumstances had also conspired to produce a situation in America where 'The listeners are concentrating their attention on local stations which give high-quality programmes and produce loud and clear signals . . . such listeners insist also upon simplicity of operation. We have, . . . also come to the end of an era of radio receivers, which are complicated to operate. . . . even more marked is the tendency towards mains operation . . . The

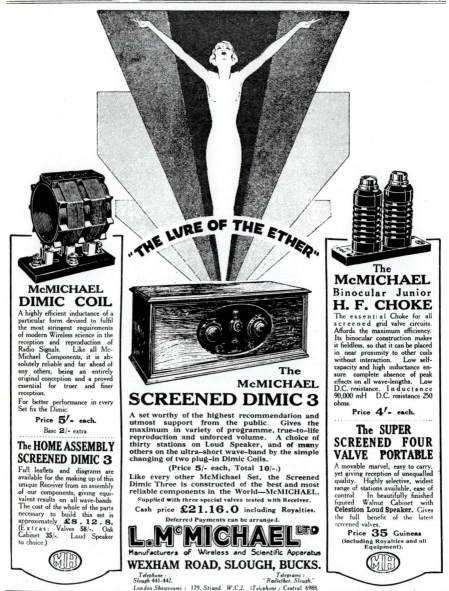

"THE LURE OF THE ETHER"

McMICHAEL DIMIC COIL

A highly efficient inductance of a particular form devised to fulfil the most stringent requirements of modern Wireless science in the reception and reproduction of Radio Signals. Like all Mc-Michael Components, it is absolutely reliable and far ahead of any others, being an entirely original conception and a proved essential for truer and finer reception.

For better performance in every Set fix the Dimic.

Price **5/-** each.

Base 2/- extra.

The HOME ASSEMBLY SCREENED DIMIC 3

Full leaflets and diagrams are available for the making up of this unique Receiver from an assembly of our components, giving equivalent results on all wave-bands. The cost of the whole of the parts necessary to build this set is approximately **£8 . 12 . 8.** (Extras: Valves 58/-. Oak Cabinet 35/-. Loud Speaker to choice.)

The McMICHAEL SCREENED DIMIC 3

A set worthy of the highest recommendation and utmost support from the public. Gives the maximum in variety of programme, true-to-life reproduction and unforced volume. A choice of thirty stations on Loud Speaker, and of many others on the ultra-short wave-band by the simple changing of two plug-in Dimic Coils.

(Price 5/- each, Total 10/-.)

Like every other McMichael Set, the Screened Dimic Three is constructed of the best and most reliable components in the World—McMICHAEL.

Supplied with three special valves tested with Receiver.

Cash price **£21.16.0** including Royalties.

Deferred Payments can be arranged.

L. M^c MICHAEL LTD

Manufacturers of Wireless and Scientific Apparatus

WEXHAM ROAD, SLOUGH, BUCKS.

Telephone :   *Telegrams :*
Slough 441-442.   " Radiether, Slough."
London Showrooms : 179, Strand. W.C.2. (Telephone : Central 6988.

The McMICHAEL Binocular Junior H. F. CHOKE

The essential Choke for all screened grid valve circuits. Affords the maximum efficiency. Its binocular construction makes it fieldless, so that it can be placed in near proximity to other coils without interaction. Low self-capacity and high inductance ensure complete absence of peak effects on all wave-lengths. Low D.C. resistance. Inductance 90,000 mH. D.C. resistance 250 ohms.

Price **4/-** each.

The SUPER SCREENED FOUR VALVE PORTABLE

A movable marvel, easy to carry, yet giving reception of unequalled quality. Highly selective, widest range of stations available, ease of control. In beautifully finished figured Walnut Cabinet with **Celestion Loud Speaker.** Gives the full benefit of the latest screened valves.

Price **35** Guineas
(including Royalties and all Equipment).

refining of . . . radio receivers has been paralleled . . . by that of gramophone pick-up equipment'.[23] An examination of the 1928 'Olympia Show Report' will show how parallel or otherwise the developments in Britain might be.

The Wireless World article 'The Trend of Progress' contains the phrases 'There can be no question that the most popular 1928–29 set is a combination of one screened grid H.F. amplifier, an ordinary triode detector . . . and a pentode output valve . . . the McMichael Screened Dimic Three, . . . is probably a pioneer of its class; . . .' '. . . complete "ganging" is, with a few exceptions, confined mainly to transportable sets which are intended to appeal to the totally unskilled, . . .' '. . . interchangeable plug in coils . . . abolished'. 'Mains-fed receivers are steadily gaining ground . . .' '. . . progress is largely bound up with valve design'. 'Accessories for the electric gramophone have reached a higher state of development . . .'.[24]

In most ways then at the end of 1928, America and Britain appear to have been running in parallel; major differences persist in the greater size, number of valves, general complexity and opulence of American sets, perhaps only to be expected in a still very wealthy country; also the British listener, with the powerful continental broadcasting stations on his doorstep, was not content with just his local station. The curiously British attachment to the portable was also an obvious difference; this, however, may be more apparent than real as the portable represented the only complete, self-contained set available to British listeners.

Moving forward to late 1929, we find that Wireless World in its 'Show Forecast' article, makes a number of very interesting statements, some of which are either paraphrased below or direct quotations given.

The point is made that manufacturers habitually kept new products shrouded in mystery until the Show opened. This commercial secrecy applied to components and, in the past, to complete receivers '. . . when these were little more than an assembly of components . . .'. In the new manufacturing conditions, however, a longer time scale was required to produce a new model so the urge for confidentiality was lessened. The article enlarges on this theme on two particular points; in particular that 'Receiving conditions have changed, . . . the introduction of the Regional Scheme and the increase in the number of stations abroad which can be listened to with enjoyment . . . make it essential that a degree of selectivity far in excess of . . . a year or so ago should be a feature of the modern set'. The screen grid valve is credited with making this possible, but at the cost of a much more exacting design and manufacturing programme necessitating a much longer time scale. The other instance given was the public demand for mains-operated equipment with an integral power supply; to this is adduced the reason why '. . . the tendency in price for complete sets of the most modern type will . . . be a little on the upward rather than the downward scale; . . .'. 'In the simpler sets, . . . prices have . . . been substantially reduced, and this season should see the eclipse of the crystal set . . . Even the B.B.C., so long the champions . . . of the crystal set user, have at last been forced to admit that under the Regional Scheme the purchase of a valve receiver is to be strongly recommended'.[25]

An instance of one of the 'complete sets of the most modern type' was given as the Marconiphone Model 47 which had an A.C. mains unit incorporated within a pressed metal cabinet and was stated to be 'a mass production job, which has made it possible to produce it at a highly competitive price'. It is, perhaps, worth noting that this set, too, bore a passing likeness to Telefunken sets of a year or so earlier. It certainly was inexpensive; at £24 it was about £15 less than the 'average' 4V mains operated receiver of its time.

In its 'Trend of Progress' article, following the show, Wireless World enlarges on the mass-production theme; for the future '. . . no very sweeping modification of generally accepted practice is anticipated. . . . many firms have deemed it safe to embark on an intensive manufacturing programme; . . . a large number of sets, . . . are designed as units, . . . This procedure – provided that production is large enough . . . makes for economy, . . . Metal lends itself admirably . . . in these modern manufacturing schemes, . . . it is almost the exception to find a receiver without a metal chassis, . . .'.[26] Reference is also made to a limited use of metal for cabinets.

It is possible that this apparent willingness to embark on the completely new manufacturing techniques referred to above and made manifest in a study of the construction of the receivers of the time, may reflect the comparative political tranquility and prosperity of the five years up to mid 1929. In which case it is perhaps as well that the process had advanced as far as it had by mid 1929; thereafter the chill winds of the Depression were to play havoc with many manufacturing industries in the country.

Two articles in Wireless World 'Receiving Sets of Today',[27] and 'Buyer's Guide',[28] enable a good picture of the market to be obtained; approximately 118 manufacturers are listed, and details of nearly 400 sets are given. Analysis of the sets shows that it is necessary to break down the groups further as specialized types of receivers were becoming more common; the figures for sets are shown in Table 5.

Table 5

Description	Battery	Mains	
Radiograms	29	49	
Superhets	8	3	
Portables	86	8	
Ordinary	119	90	
Totals	242	150	= 392

Hence 150, some 38%, of all receivers, were mains operated. If superhets and portables are excluded, there were 148 battery sets and 139 mains sets. It should be stated here, however, that what is classified as a 'mains' receiver is somewhat arbitrary; some are no more than a battery set with an eliminator of some sort in the battery compartment. Also quite a few were for use only on D.C. mains, with ordinary filament valves supplied through a dropping resistor. Nevertheless, 72 sets are shown as using indirectly heated valves.

In order to compare with previous years, it was thought better to exclude superhets and radiograms, the former were all lingering antiquated designs, and radiograms comprised several classes in themselves; some were just pick-up power amplifiers with a simple tuning circuit added, others were a chassis from a set dealt with elsewhere united with a turntable in a more-or-less exotic cabinet.

With these exclusions, then, the breakdown is given in Table 6.

Table 6

1929	Portable			Ordinary			Mains			Totals		
Valves	No.	%	Price	No.	%	Price	No.	%	Price	No.	%	Price
1V	2	2	£ 7	1	1	£ 7	–	–	–	3	1	
2V	5	6	£13	21	18	£12	17	19	£17	43	14	
3V	3	3	£15	57	48	£14	45	50	£27	105	36	
4V	29	34	£23	23	19	£30	21	23	£39	73	25	
5V	47	55	£23	17	14	£26	7	8	£48	71	24	
Over 5V	–	–	–	–	–	–	–	–	–	–	–	
Totals	86	30		119	40		90	30		295		

Within the table several remarkable features arise; notably the rise in the percentage of mains sets, mirrored by the decline in both battery types; even if the problems of A.C. mains could not be regarded as fully solved, (new valve developments would later attend to that), the end had clearly started. In the portable sets, there is a clear division between the older designs, primarily 5 valve all triode, and the newer circuits with multi-electrode valves. This appears to be the explanation for the identity of price between 4V and 5V sets (a difference more apparent than real, there is a 'rounding' process applied to calculating average prices which tends to group them into batches). Similar features are seen in the ordinary battery receivers, where the average 4V set was much more expensive than a 5V one! Again, two types of circuit were elbowing each other aside in this transition period; those with the greater number of valves principally arise from more or less obsolescent designs.

Compared with 1928 figures, prices had indeed edged upwards, but as 'value for money' the newer sets were much better bargains. One trend which has already been remarked upon was the clear establishment of an S.G.-detector-pentode combination as the most favoured design for the most recent sets. The quite spectacular growth in numbers of radiograms is noteworthy.

The 'Receiving Sets of Today' article stresses other design features which are not apparent above. Almost all H.F. amplification is shown to be by S.G. valves; just over half the sets now have 'ganged' (or semi 'ganged') tuning, about one quarter no longer used reaction and just over one quarter of current production sets had metal cabinets.

Moving on to 1930, there are again approximately 118 manufacturers listed and again nearly 400 receivers.[29] The names of the manufacturers, however, changed as firms disappeared from the scene and others moved in; familiar names from the earliest days, like Henderson and Tangent are missing. The Wireless and Electrical Trader revealed that 649,000 sets of all types, excluding kits, were produced in 1930 and that a typical 3 valve battery set sold at £4.10.0, as against £20 in 1923.[30] The licence curve at the beginning of this chapter suggested 900,000 new licences in the year, so something over 250,000 sets in kit form would be a reasonable estimate. The prices given seem low.

The 'Show Forecast' uses one word 'consolidation' to sum-up the characteristics of the scene in 1930; little more than detail improvements were to be expected. There is reference to the position of the Marconi Royalty, recently reduced, which was encouraging a more generous use of valves; also to the reduction in the number of portables as the obsolete 5 valve aperiodically coupled triode circuit was finally laid to rest. In a passing reference which has a bearing on the political situation, the article states: 'The British manufacturers intend to make a bold bid to show the public that there is no need to go beyond the shores of our own country to satisfy requirements . . .'.[31] The 'Trend of Progress' article reinforces the picture of consolidation, with worthwhile improvements to many receivers but no great innovations. Reference to Transmitter policy is made, as simple (2 valve) sets suitable for 'Regional Twins' (even fixed tuned in some cases), could be obtained and because of mass-production methods, at very low prices.[32]

A detailed analysis of the sets briefly described in the 'Buyer's Guide'[33] seems scarcely worthwhile in view of the lack of major change although much information is in any case contained in an accompanying article;[34] it is also beginning to be difficult to compare conditions with those obtaining back at the beginning of broadcasting, as sets became much more specialized. A few statistics would help to highlight the picture: 393 sets were listed; 179 battery operated; 172 with I.H. valves and 42 designed for D.C. mains – or using some other form of battery eliminator – thus the mains sets for the first time outnumbered battery sets. Portables show a decline, from 86 to 68 over the year, but this was more or less exclusively caused by the withdrawal of outmoded designs; radiograms and record-players show some small advance. In most cases, prices for the same sets had not changed greatly, sometimes a slight reduction – perhaps by 'pounds' instead of 'guineas'. The following Manchester Show had little new to report, with the major exception that '. . . Cossor must not only be looked upon as producers of valves, but also as manufacturers of a complete range of modern receiving sets'.[35]

Wireless World gives an account of the New York Show; there appears to have been very little change over the year, although there is a suggestion that superhets, now incorporating S.G. valves, are making a reappearance.[36] By contrast, the review of the Paris Show makes the point that, because of American influence, the ubiquitous superhet now has a rival in the multi H.F. stage receiver based on American designs.[37] Possibly because of financial problems, possibly just to draw breath, the world-wide wireless industry appears to be pausing. One feature, to which passing reference has been made, is the change in scale of the manufacturing industry. A distinct shift from small, independent constructors, or even assemblers, to large firms already having major manufacturing capacity – perhaps in another field – is evident.

Gone are the 'Alfred Pearson, 80A Newland Avenue, Hull' type of firm, who can never have been more than "constructors" rather than manufacturers; in come the 'Ekco', 'Murphy'. 'Ferranti' 'Cossor' names, soon to be followed by others such as 'Mullard', firms of already considerable scale and with experience in mass-production methods as well as the resources to finance a complete revolution in manufacturing techniques in the wireless industry. Their arrival on the scene must have caused considerable alarm in the board rooms of the few large-scale companies operating up to that time – Marconiphone, G.E.C., etc. In a way, this time of flux is reminiscent, although on a different scale, of the situation at the start of broadcasting, when cycle dealers, garages, lamp-makers diverted into the new craze, and even considerable existing concerns, like A.J.S., found it a profitable sideline. Now that wireless had come of age, ceased to be a passing craze and settled down into nationwide acceptance as a necessity; now that alternative programmes were in the air and programme content was of more import that just the miracle of hearing; so a rebirth of the manufacturing industry on an increased scale to supply the increased requirements of this new situation was due, perhaps overdue. The time was ripe in other ways as well; politically and economically turbulent times were at hand but, with the National Coalition Government of August 1931, relative tranquility returned to the domestic political scene. Even the financial retrenchment and straitened economic circumstances which followed were more suited to the survival of the large scale efficient industrial process than the two preceding, uncertain, years had been. Certainly the economic squeeze hastened the demise of many small concerns. Their bigger brothers, however, were able to make good use of the market opportunities thus presented, helped also by the protective tariff introduced in 1931, and the British wireless industry set about developing its home market with a will. As we have seen, the potential market for receivers was set to grow.

The British industry had clearly been open to ideas on design and mass-production methods from overseas notably from America and Germany; some of these influences did not persist, for instance, metal 'cabinets', (probably having their origin on the Continent), had only a brief airing here, whilst moulded plastics, also of Continental origin, became very well established. On the whole, British manufacturers, with the outstanding exception of valve-makers, seemed to be rather conservative in their new designs; perhaps the head- start of America in the matter of broadcasting meant that they were of necessity in the shadow of new developments from there, but they were sensitive to these new ideas; perhaps the dire warnings made manifest by the entirely insular – and backward – French wireless scene made a telling point.

Whatever the significance of all these influences, the fact remains that the 1929–32 era saw the birth of the modern British wireless industry. With a very advanced valve-making industry at their back, it was able to move into the new era with a quite remarkable rate of progress.

Turning-point

The year 1931 seems to have been the turning-point. Then so many of the background influences which had long been at work finally coalesced and led to the receiver design which, almost unbelievably rapidly, came to dominate production continuing even to the present day. The modern superhet was born.

We have already looked at a few of the technical aspects of this concept; sufficient to recall here that the advanced S.G. valve, the new capability for mass manufacturing components to close tolerances, the lessons and techniques learned for efficient screening of R.F. amplifiers, these were some of the essential prerequisites for the new superhet which made it a very different animal from its highly temperamental predecessor. The new design was indeed an elegant solution to the seemingly insoluble twin problems of selectivity with quality and sensitivity without multiplicity of valves.

The early stages of this revolution, for such it was, are difficult to trace. Superhets in very limited numbers did appear at the New York Show of 1930, even then in typical American fashion using 12 or 14 valves; but the patent position in the U.S.A. appears to have been very confused.[38] An article in Wireless World on 'Recent Developments in America' states in reference to superhets, '. . . the patent situation has been definitely cleared up and the R.C.A. has the monopoly of superheterodyne patents'. R.C.A. was, apparently, prepared to license other manufacturers, but only at a price. The article continues 'The R.C.A. has recommenced the manufacture of an improved model', for which a very full description is given, including the telling phrase '. . . the degree of selectivity is so high, . . . that a considerable amount of sideband cutting takes place'.[39] This was one of the undesirable characteristics of the old form of superhet, but it is just possible that here it was deliberate, as the comparatively recent introduction of moving coil loudspeakers at last allowed real bass response to be provided and American up-to-date receivers were expected to provide it in no half-hearted manner.

From this evidence it appears that American superhet development had been hampered and delayed by the litigation involved in disentangling the patent position and the financial climate, too, would call for caution. Around this time Wireless World itself was promoting the cause of the superhet, a fact of which it constantly reminded its readers in later months. A. L. M. Sowerby had an article giving a very detailed discussion of the

MARCH 18th, 1931.   Wireless World   293

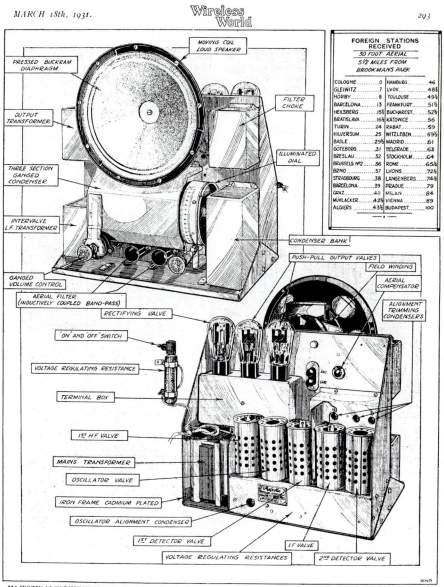

FOREIGN STATIONS RECEIVED
30 FOOT AERIAL
5½ MILES FROM BROOKMANS PARK

COLOGNE	0	HAMBURG	46
GLEIWITZ	7	LVOV	48½
HÖRBY	8	TOULOUSE	49½
BARCELONA	13	FRANKFURT	51½
HEILSBERG	15½	BUCHAREST	52½
BRATISLAVA	16½	KATOWICE	56
TURIN	24	RABAT	59
HILVERSUM	25	WITZLEBEN	59½
BASLE	29½	MADRID	61
GÖTEBORG	31	BELGRADE	63
BRESLAU	32	STOCKHOLM	64
BRUSSELS N°2	36	ROME	65½
BRNO	37	LYONS	72½
STRASBOURG	38	LANGENBERG	74½
BARCELONA	39	PRAGUE	79
GRAZ	40	MILAN	84
MUHLACKER	42½	VIENNA	89
ALGIERS	43½	BUDAPEST	100

Labels (diagram):
MOVING COIL LOUD SPEAKER
PRESSED BUCKRAM DIAPHRAGM
FILTER CHOKE
OUTPUT TRANSFORMER
ILLUMINATED DIAL
THREE SECTION GANGED CONDENSER
INTERVALVE L.F. TRANSFORMER
CONDENSER BANK
GANGED VOLUME CONTROL
PUSH-PULL OUTPUT VALVES
FIELD WINDING
AERIAL COMPENSATOR
ALIGNMENT TRIMMING CONDENSERS
AERIAL FILTER (INDUCTIVELY COUPLED BAND-PASS)
RECTIFYING VALVE
ON AND OFF SWITCH
VOLTAGE REGULATING RESISTANCE
TERMINAL BOX
1ST H.F. VALVE
MAINS TRANSFORMER
OSCILLATOR VALVE
IRON FRAME CADMIUM PLATED
OSCILLATOR ALIGNMENT CONDENSER
1ST DETECTOR VALVE
I.F. VALVE
VOLTAGE REGULATING RESISTANCES
2ND DETECTOR VALVE

HIND

MAJESTIC SUPERHETERODYNE. Tuning positions on the single dial control are given for the reception of the principal foreign stations.

A 25

whole basis of the superhet and showed it to be fully capable of providing the selectivity now considered essential. He makes the very clear statement '. . . it should now be possible to design a superhet which would be free from all the defects commonly attributed to receivers of its type', and again 'While the writer does not anticipate that the present type of receiver will need to be superseded for short-range work, he is inclined to think that for long distance reception the superhet, in some form or other, will eventually return to its old position of dominance'.[40] In the article he clearly advocates an I.F. of the order of 100 kc/s for a modern superhet.

Just a month later appeared a remarkably comprehensive design for a 6 valve superhet, attributed to A. L. M. Sowerby and H. B. Dent, and intended for home construction.[41] It had a number of curious features which suggest that the old form of superhet had not been entirely abrogated; a frame aerial was still specified (although the reasons requiring its adoption were different) and the very old-fashioned I.F. of 30 kc/s was adopted. Undoubtedly the worst of the old features still retained was the need for no less than three tuning dials. In spite of the claims, this was no modern superhet, the only advance shown was in the use of S.G. valves, and even these were not necessary at 30 kc/s.

Then came a most odd and puzzling incident: on 18 March 1931, Wireless World gives a full, illustrated, description of a 'Majestic Model 50 Screen Grid Superheterodyne'; a 7 valve + rectifier compact mains operated receiver with built-in loudspeaker which was in every way the prototype model of the high quality complete domestic wireless-set which appeared in its millions throughout the 1930s in every corner of the U.K. The article recognizes the significance of this set, it concludes 'Here we have a receiver which, complete with its nine valves (mistake) and selling at 28 guineas, probably represents a forerunner of both the superheterodyne and the compact totally self-contained types'. The receiver was attributed to the 'Majestic Electric Co. Ltd., Majestic Works, Tottenham, London, N17',[42] but the entire specification, external appearance, even the shape of the valves, cries 'American'. No reference is made to this. The set was obviously built like a battleship on an iron frame with cadmium plated heavy-iron compartments for screening. Single knob tuning was achieved by the use of padding and trimming capacitors in the oscillator tuned circuit, those points that the earlier Wireless World article maintained were very difficult to attain but which later became more or less standard. Another pointer to American practice was the omission of the long wave band; indeed the choice of I.F., 175 kc/s (1700 m) rendered it impossible; also the article remarked on the attenuation of the higher audio frequencies, an American characteristic. The matter was put beyond reasonable doubt, however, as a beautifully drawn diagram of the chassis of the set bore very clearly the legends 'ANT' (for aerial) and 'GND' (for earth). Furthermore a peculiar voltage dropping resistor is shown fitted in the mains lead by the on/off switch at the side of the cabinet; this would be necessary to drop the British mains voltage of up to 250 volts to fit the standard American supply of 115V. Everything suggests an American receiver slightly modified for the British market. Was this one of the 'one or two superhets' reported as being on display at the New York Show of 1930? There is one very curious feature; as part of the comprehensive smoothing for the H.T. supply, a 'tuned rejector' was created by connecting a 0.09 μF capacitor across a 30H choke. The response curve of this would be very broad, but even so such a circuit would resonate at 98 c/s and a long way from the 120 c/s required for America; a capacitor of about 0.06 μF would be needed for this purpose. This indicates some forethought in adapting to British conditions.

It seems most improbable that a hitherto unknown British company could suddenly appear on the market with a fully developed modern superheterodyne many months before anybody else. The Manchester Show report, six months earlier, gives the same set as on display by 'Majestic Distributors (Manchester) Ltd., 2 Victoria Bridge Street, Manchester' it states clearly that 'The Majestic receivers . . . distributed in the North of England by this firm, are of interest not merely because they are of American manufacture, . . .'.[43] The critical feature appears to have been that the Manchester Show was organised by the 'Evening Chronicle' and not the Radio Manufacturers Association (R.M.A.), so foreign sets could be shown. The R.M.A. did not make the same mistake the following year. This set must therefore have been produced in America, possibly illegally, and shipped here as a way of trying out a new design ready for instant sale in the U.S. when the patent position was clarified. Notably, other foreign exhibits were present at this one show.

Wireless World ascribes to itself much of the growing interest in the superhet 'It is not too much to say that, since the revival of the superheterodyne by this journal in October last, it is likely to continue in favour until such time as a means is devised of giving the tuned H.F. set the same high selectivity'.[44] It publishes a second circuit for home constructors '. . . its construction is considerably easier than that of a straight H.F. set, . . .';[45] but this 5 valve + rectifier A.C. mains set would tax the abilities of even an experienced constructor; the baseboard was packed with components. The I.F. had been moved up to 110 kc/s, so it was progressing towards a modern superhet, but it still required two tuning controls. In the conclusion of the article it was stated that a powerful station may be expected to have four separate tuning points! A battery version followed looking only a little less complex and consuming a massive 20mA anode current from the H.T. battery.[46] Certainly these were not sets for the faint hearted. Ironically, they illustrate many of the problems which had made the old superhets so unpopular.

Turning to commercial receivers, an editorial predicts that in the new sets to be expected at the Radio Show in September '. . . the majority of the manufacturers will have standardized a straight H.F. set with all modern refinements, though a few superheterodynes will make their first appearance'.[47] The reasoning behind this assertion is that the two dial tuning system for superhets will not be popular with the public. However, in the 'Trend of Progress' article, after the show occurs the phrase '. . . the superheterodyne . . . has come right to the front'.[48] Single dial tuning, metal chassis punched and bent into shape and moving coil speakers are also mentioned as new features, reflecting the new mass-production capabilities. The same enthusiasm for the new superhet was shown in the section on 'New Receiver Designs' where superhets were discussed before all other sets. It is clear that design was far from stable as yet and the article states '. . . that many widely different circuits are employed . . .', and in spite of their own reservations 'In many cases, also, the oscillator condenser is "ganged" to the others, thus giving a true single control of the receiver'.[49] In contradistinction to their own design, mention is made of the Edison Bell 5 valve battery superhet which consumed only 9mA H.T. current, good value at 20 guineas. It is clear that some designers had come under influences other than American; the Halcyon Transportable had a typical French 'bigrille' (two grid) valve type as frequency changer, although it was a Mullard PMIDG (a type of which Wireless World was not fond).

In such a state of flux, it would not be useful at this stage to do a complete analysis of the 1931 set designs, beyond mentioning that '. . . the new sets are offered at extraordinary low prices'. The Wireless and Electrical Trader revealed that a grand total of 1,258,197 sets had been produced in 1931. The average retail profit was 33.33% at this time, as compared to 25% in 1923.[50]

The 1932 pre-show article of Wireless World, 'What to see at the Show', makes the statement 'The superheterodyne forms one of the most interesting classes of receiver this year, and . . . few firms . . . are not showing at least one model; a number have abandoned the straight receiver almost entirely . . .'.[51] After the Show, in the 'Trend of Progress' article it is moved to record 'Receiver design has reached . . . a state of stability. . . The tendency towards a self-contained set, with a built-in loud speaker, . . . has now become established practice, . . . the three-valve H.F.-det.-L.F. set maintains its ascendancy, but the superheterodyne continues to gain ground in the long-range field, . . . even the cheaper sets have seldom more than one tuning control'.[52]

These quotations illustrate the point made in full in the article and clearly shown in a study of the stand-by-stand show reports, that the superhet was very far from attaining a position of complete ascendancy. Some manufacturers, notably Philips, had firmly resisted the trend by instead producing new T.R.F. sets using the new valves and equipping them with very high Q (superinductance, 'Litz' wire on glass formers) coils to give enhanced selectivity. By omitting reaction from these sets and introducing 'band-pass' tuning a considerable degree of selectivity was still possible without unduly restricting the upper audio register. Receivers were available over a whole spectrum of cost. At one end were two valve det.-L.F. sets, also common in Germany at this time, both battery and mains, some of which reached quite a high level of sophistication. The Pye 'K'[53] with band-pass tuning gave a good account of itself and at 12 guineas was a relatively inexpensive answer for those who were perhaps concerned with local station listening. At the other end of the scale there was considerable interest in expensive equipment, often radiograms. The H.M.V. Company produced models incorporating 7 – or even 10 – valve superhets, and retailing at prices up to 80 guineas.

Wireless World, July 22nd, 1932.   *63*

A 2-VALVE SET WITH BAND-PASS INPUT CIRCUIT.

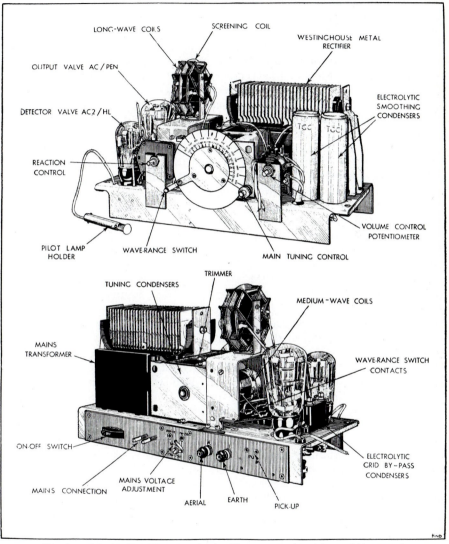

Two views of the Pye Model K receiver chassis.   The power transformer is completely screened.

Some new valves were available; metallized forms of current type cut down on expensive screening and the variable mu S.G. valves had allowed the elimination of 'cross-modulation' yet another of the superhet's problems, but further circuit development in Britain was, for the first time in recent years, being hampered by the unavailability of valve varieties which were to become very common in years to come. Notable examples of missing types were single valve frequency changers and double-diodes for second detectors, although Murphy at least, as a stop-gap, had pressed into service a full wave mains rectifier in their A8 superhet ('. . . "duo-diode" . . . an ordinary Mazda indirectly heated full-wave mains rectifier type UU2 . . .').[54] In respect of special valve types, the Americans were ahead (Wunderlich valve)[55] and thus the full potential of the superhet was not being realized in Britain. W. T. Cocking wrote in Wireless World 'Little published data has appeared in this country on the subject of automatic volume control, and at the time of writing it is believed that no British set includes this refinement. Such controls are quite common in America, . . .'[56] he does not come out unequivocally in favour of the system.

The position at the end of 1932 was clearly expressed in a Wireless World editorial '. . . this country has by no means recovered from the depression, . . .' but it regards the figures for business done at the recent show as '. . . extremely encouraging . . .'.[57] These figures showed that the industry as a whole expected to sell £50M worth of radio instruments and components (2M sets and 10M valves) to the home market in 1933, compared with only £29M the previous year. Figures such as these certainly show the renewed business confidence that would be expected from the more settled political and economic conditions within the country following the re-election of the National Government. The prediction was well above the actual figure, however, as shown by the figures given in The Wireless and Electrical Trader; referring to 1933, here it stated '. . . a further statistical survey showing that 967,000 sets had been produced the previous year'.[58] From the graph we find 900,000 new licences were taken out in the year; this would not represent the number of sets bought, however, as the home constructor especially of kit sets, would be considerable. To balance this, there was an increasing need – and opportunity – to replace obsolete equipment, so sales close to the new licence total could perhaps be expected.

Many of the valve deficiencies noted last year, were remedied during 1933; the Show had a crop of 'heptodes', 'double-diode triodes' etc. The progress of set design continued unabated throughout the year, generally prices fell, as did the numbers of valves in the 'average' receiver, largely due to the new multiple varieties. A Wireless World editorial expressed unease at the attempt by manufacturers to bring prices down to '. . . a dangerous figure'[59] and The Wireless and Electrical Trader revealed that 'Too many home-produced models were . . . in a faulty condition and prone to unreliability in service'.[60] A year later the 'New Receiver Designs' article published in anticipation of the show, asserts '. . . the "standard" set for the 1934-35 season is to be . . . a small superheterodyne, having as a rule a total of four valves'. On the subject of costs 'The typical small mains superheterodyne now costs roughly 12 guineas for the table model'. A.V.C. was stated to be included 'almost as a matter of course in all super-heterodynes' and 'moving-coil speakers will be fitted in almost every set costing more than five or six pounds'. More or less the complete range of valves was now available, 'universal' or A.C/D.C. sets were appearing, straight sets were mentioned '. . . the designers . . . have ceased to compete with the superhet'[61] and this in spite of the general availability of new magnetic materials for producing very high Q coils.

It is clear that over the decade 1924-34, by far the greater change took place in the second half. Receivers transformed from bulky, exposed apparatus attached to numerous essential accessories to become complete self-contained, largely mains-operated units. Circuits evolved from straight sets with H.F. stages, (if present), of doubtful value, fitted with multiple tuning knobs, obligatory reaction, and unpleasant habits, to become docile, well behaved superhets with enormous potential I.F. gain well held in check by A.V.C. Loudspeakers progressed from 'phones on the table' to having attached ear-trumpets, to beautiful looking! swan-neck horns, through 'cones' of every conceivable form to the standard moving-coil-diaphragm, with us still. As receivers became simpler to operate, and circuits refined down largely to one standard type, valves went the other way, from simple elementary triodes of perhaps thousands of allegedly different types in hundreds of different circuits, to a genuine variety of completely different valves, performing completely different functions but in good circuit design co-operating in harmony. Truly a transformation!

Subsequent developments were numerous; for a time perhaps rather 'cosmetic', with new 'furniture' designs, great clear tuning dials, tuning aids, tone controls, all-wave and short-wave (S.W.) working. This last merged into television – but that is another story.

It is perhaps enough to summarize the position in a few figures. The table presented repeats some of the information more fully discussed earlier along with a brief analysis of data from later years. Figures are not directly comparable in all cases; in particular as receivers developed and became more specialized it becomes difficult to relate prices.

In 1934 the scale of some of the firms became apparent; over one third of all sets offered for sale came from only eight manufacturers. In other cases, some smaller companies offered only one or two – often expensive – highly specialized designs. There was a tendency, too, for the same chassis to appear in different cabinets under different guises, sometimes attached to a turntable as a radiogram. Such practices conceal to some extent the great measure of standardized mass-production, as well as disguising the remarkably low prices. In 1924, £12 would have bought no more than a single valve set and accessories, requiring constant attention to all its battery supplies, the problems of a full-scale outdoor aerial and the inconvenience of headphones; by 1934 the same sum would have given a choice of several first class mains operated receivers capable of picking up most European broadcasting with no external aerial, all at the 'flick of a switch'.

Brandes
The Name to Know in Radio.

*Result of
16 years'
experience*

Brandes and the Woman.

The male species often find occasion to bewail the erratic tendencies and the inconsistency of the opposite sex. As far as Brandes are concerned we disagree entirely. They have always appealed to the woman. She has had a keen discernment for the value of these famous Headphones. Whilst not entirely conversant with what " Matched Tone " means, she is thoroughly convinced of their sweet toned reception and absolute comfort. The featherweight headband does not disturb or catch in the hair—get Brandes from your dealer.

Manufactured at Slough, Bucks, by
Brandes Limited, Walmar House, 296, Regent Street, London, W.1.
Glasgow—47, Waterloo Street.   Newcastle—5/6, Post Office Chambers.

PRICE
25/-

*British manu-
facture (B.B.C.
Stamped) and
conform to all
new licensing
regulations*

*'Phone—Mayfair
4 2 0 8 - 4 2 0 9.*

*Trade enquiries
invited.*

Matched Tone
TRADE   MARK
Radio Headphones

The WORLDS STANDARD WIRELESS LOUD SPEAKER

The AMPLION JUNIOR with Floating Diaphram **27/6**

The AMPLION JUNIOR DE LUXE with Floating Diaphram **£2:2:0**

IN USE IN · G⁺ BRITAIN · IRELAND · NORWAY · SWEDEN · FRANCE · HOLLAND · ITALY · BELGIUM

IN USE IN · AMERICA · AUSTRALIA · N. ZEALAND · S. AFRICA · INDIA · DENMARK · SPAIN · JAPAN

Price Reductions.

THESE models incorporate the latest developments in Loud Speaker construction. . . . Colourable imitations of the above original designs having been placed upon the market . . but without the patented features exclusive to the Amplion and essential to Loud, Clear and truly Natural reproduction . . the House of Graham has decided to protect the public by marketing genuine Amplion models at these greatly reduced prices.

ALFRED GRAHAM & COMPANY
(E. A. GRAHAM)
St. Andrew's Works, Crofton Park,
LONDON, S.E. 4.

Telephone ·
Sydenham 2820-1-2
Telegrams :
"Navalhads.,
Catgreen, London."

Showrooms :
25-6, Savile Row,
W.1., and 82, High
St., Clapham, S.W.s

AMPLION

In replying to advertisers, use COUPON on last page

iv. ADVERTISEMENTS.   THE WIRELESS WORLD AND RADIO REVIEW   NOVEMBER 24TH, 1926.

"Type **33**"
Mahogany or
walnut finish.
£5 · 5 · 0.

STERLING
"TYPE 33" LOUD SPEAKER

THIS model has been designed primarily to improve the reproduction of the lower notes without sacrificing the higher tones, in order to effect a more faithful rendering of broadcast music than has heretofore been possible with commercial loud speakers.

In the past all efforts to successfully reproduce the low tones, whether by horn or cone, have resulted in a general falling off in the efficiency of the instrument, thus increasing the number of valves required in the receiving set. The volume of output for a given input will be found to be greater than with the majority of loud speakers, while the range is far greater. From the pedal notes of the organ to the top notes of the piccolo the music is reproduced with great faithfulness. Height 23″, diameter of flare $14\frac{3}{4}$″.

Send for literature describing the full range of Sterling Loud Speakers.

THE STERLING "DINKIE" LOUD SPEAKER—*the little fellow with the big voice.*
£1 . 10 . 0.

The Marconi International Marine Communication Co., Ltd., require 500 qualified wireless operators. Apply to Service Manager, Marconi House, Strand, London, W.C.2.

THE MARCONIPHONE COMPANY LIMITED.
(Sole Agents for the Sterling Telephone and Electric Co. Ltd.)

Head Office: 210-212, Tottenham Court Road, London, W.1.

Regd. Office : Marconi House, Strand, London, W.C.2.

Printed for the Publishers, ILIFFE & SONS LTD., Dorset House, Tudor Street, London, E.C.4, by The Cornwall Press Ltd., Paris Garden, Stamford Street, London, S.K.1. Colonial and Foreign Agents :

14   ADVERTISEMENTS.   THE WIRELESS WORLD   JUNE 6TH, 1928.

Leadership since 1911—
MAGNAVOX
The Originators of the Moving Coil Speaker

Radio's first loud speaker—still the finest

The moving coil type of loud speaker was created by Magnavox in 1911. It is the only type of speaker that has stood through every period of speaker development. Supreme in the beginning—supreme to-day. Protected and controlled by Magnavox exclusive patents. There are nearly half a million Magnavox Dynamic Speakers now in use.

MODELS FOR A.C. OR D.C.

TYPE D.80/1. For operation from A.C. Lighting Mains 100/120 volt 50/60 cycle A.C.
£11-11-0

„ D.80/20. For operation from A.C. Lighting Mains 200/240 volt 50/60 cycle A.C.
£11-11-0

„ R.5. For operation from D.C. Lighting Mains 100/240 volt D.C.
£10-10-0

„ R.4. For operation from 6 volt accumulator or Trickle Charger. Consumption ½ amp. **£9-10-0**

THE "GREAT VOICE"

This new 32 page booklet is now ready. Tells you all about moving coil speakers, power and gramophone amplifiers, volume controls, pick-ups, etc. Numerous circuits and illustrations will enable you to get the finest performance from your Magnavox Unit. This indispensable booklet will be sent you on receipt of 6d. to cover cost and postage. Write for your copy immediately.

Telephone:- Mayfair 0578-9

The ROTHERMEL CORPORATION . LTD.
24-26, MADDOX ST, LONDON, W.1.

Telegrams:- Rothermel, Wesdo, London

Table 7

Year	1924	1926	1928	1929	1930	1932	1934
No. of Manufacturers		64	107	118	118	79	43
Sets %		17% / 83%	30% / 70%		44% / 16% / 40%	47% / 24% / 29%	49% / 18% / 33%
Sets No.		59 / 297	90 / 205		145 / 75 / 309	89 / 309	110 / 42 / 74 / 226

Types — valve distribution

Valves	1926 No.	1926 %	1928 Mains	1928 Batt	1928 Total	1928 %	1929 Mains	1929 Batt	1929 Total	1929 %	1930 AC	1930 DC	1930 Batt	1930 Total	1930 %	1932 AC	1932 DC	1932 Batt	1932 Total	1932 %	1934 AC	1934 DC	1934 Batt	1934 Total	1934 %
1	41	9	–	3	3	1	–	3	3	1	–	–	1	1	0.3	–	–	–	–	–	–	–	–	–	–
2	139	30	12	68	80	22	17	26	43	14					13	14	10	18	42	3	3	1	5	9	4
3	126	27	27	105	132	37	45	60	105	36					39	51	41	43	135	44	19	8	38	65	29
4	128	27	13	36	49	14	21	52	73	25					30	38	12	19	69	22	49	16	20	85	38
5			7	70	77	22	7	64	71	24					14	20	7	5	32	10	15	5	10	30	13
6	36	7	15	15		4	7								3.7	9	2	4	15	5	10	6	5	21	9
7	}	}	}	}	}	}	}	}	}	}	}	}	}	}	}	10	3	–	13	4	5	3	2	–	2
Over 7	}	}	}	}	}	}	}	}	}	}	}	}	}	}	}	3	–	–	3	1	11	–	–	11	5
No.	470																		309					226	

	1924	1926	1928	1929	1930	1932	1934
Average Price	£26	£25	£29 / £20 / £22	£30 / £20 / £23	£30 / £24 / £23	£10 / £20	£22 / £17 / £11 / £17
No. of Superhets						52	140
Percentage Superhets						17%	62%

References

1 Daily Mail Year Book, 1927, p. v
2 Daily Mail Year Book, 1927, pp. 20ff
3 Wireless World, 26 May, 1926, back cover
4 The History of Broadcasting in the United Kingdom, Vol. I, p. 358
5 Wireless World, 24 November 1926, p. 704
6 Wireless World, 12 January 1927, p. 53
7 Wireless World, 10 February 1926, pp. 199ff
8 Wireless World, 3 issues in December 1934
9 Wireless World, 10 February 1926, pp. 199ff
10 Wireless World, 5 October 1927, pp. 478ff
11 Wireless World, 21 September 1927, pp. 369ff
12 Wireless World, 31 August 1927, pp. 267ff
13 Wireless World, 5 October 1927, p. 483
14 Wireless World, 9 February 1927, back cover
15 Wireless World, 5 October 1927, pp. 478ff
16 Wireless World, 12 October 1927, pp. 513ff
17 Wireless World, 27 June 1928, pp. 685ff
18 Wireless World, 15 September 1926, p. 388
19 Wireless World, 14 November 1928, pp. 654ff
20 Wireless World, 14 November 1928, pp. 663ff
21 Wireless World, 21 November 1928, p. 706
22 Wireless World, 14 September 1927, p. 339
23 Wireless World, 27 June 1928, pp. 685ff
24 Wireless World, 3 October 1928, pp. 461ff
25 Wireless World, 18 September 1929, pp. 269ff
26 Wireless World, 2 October 1929, pp. 369ff
27 Wireless World, 20 November 1929, pp. 552ff
28 Wireless World, 20 November 1929, pp. 560ff
29 Wireless World, 19 November 1930, pp. 576ff
30 The Wireless and Electrical Trader, 11 March 1944, p. 297
31 Wireless World, 17 September 1930, pp. 277ff
32 Wireless World, 1 October 1930, pp. 377ff
33 Wireless World, 19 November 1930, pp. 576ff
34 Wireless World, 19 November 1930, pp. 572ff
35 Wireless World, 15 October 1930, pp. 433ff
36 Wireless World, 29 October 1930, pp. 492ff
37 Wireless World, 22 October 1930, pp. 463ff
38 Wireless World, 29 October 1930, pp. 492ff
39 Wireless World, 11 February 1931, pp. 138ff
40 Wireless World, 1 October 1930, pp. 393ff
41 Wireless World, 5 November 1930, pp. 512ff
42 Wireless World, 18 March 1931, pp. 292ff
43 Wireless World, 15 October 1930, p. 439
44 Wireless World, 13 May, 1931, pp. 498ff
45 Wireless World, 3 June 1931, pp. 597ff
46 Wireless World, 15 July 1931, pp. 59ff
47 Wireless World, 29 July 1931, editorial
48 Wireless World, 30 September 1931, p. 381
49 Wireless World, 30 September 1931, p. 382
50 The Wireless and Electrical Trader, 1 March 1944, p. 297
51 Wireless World, 19 August 1932, p. 156
52 Wireless World, 2 September 1932, p. 231
53 Wireless World, 22 July 1932, pp. 62ff
54 Wireless World, 2 September 1932, p. 244
55 Wireless World, 23 September 1932, pp. 290ff
56 Wireless World, 12 August 1932, pp. 116ff

57 Wireless World, 9 September 1932, editorial
58 The Wireless and Electrical Trader, 25 March 1944, p. 343
59 Wireless World, 25 August 1933, editorial
60 The Wireless and Electrical Trader, 25 March 1944, p. 343
61 Wireless World, 10 August 1934, pp. 94ff

Chapter 7
Home construction and kit sets

From the earliest days of wireless, amateurs had played a considerable part in its development. According to an address to the R.S.G.B. by the President, Dr. W. H. Eccles, F.R.S., in 1923, on 'The Amateur's Part in Wireless Development'. '. . . many of the great advances in wireless have been initiated by the amateur, and that most of the early steps in the inception of the subject were taken under the stimulus and guidance of men who were neither telegraphists or engineers, but merely lovers of the infant science'.[1] His definition of 'amateur' is narrow; most workers in the field were amateurs necessarily as it was not an established study.

An interesting article concerning British pre-broadcasting days occurs in Wireless World. The writer, commenting on the paucity of shops selling radio components, instances journeys from the Midlands to London or even Manchester to find insulated wire, and indicates that valves sent through the post were prone to come to grief. He continues '. . . Ordinary component parts were usually taken from ex-Government sets, practically everything was second-hand . . . generally, you made everything possible yourself.' 'The erection of an aerial set the district guessing; . . . You were regarded as something of a marvel; . . .'. He concludes on a wistful note '. . . We are told that we are rapidly approaching an era of international broadcasting, but I think some of the glamour of the old days has gone'.[2] A comment on some of these practices is given in an American book of the time, 'The Home Radio: How to Make and Use it' '. . . anyone with the least mechanical ability can build wireless telephone sets if they purchase the parts which require special knowledge, skill or devices for making.' '. . . a variable grid-leak can be made with pencil marks on paper by erasing or adding lines . . .' '. . . it does not pay to try to make certain instruments . . .'.[3] Further informative comment appears in a Wireless World editorial, 'Popularising Wireless is usually regarded as making the subject simple and devoid of all theory, but anyone who regards himself as an amateur or experimenter would never be satisfied to disregard a study of theory, however superficial, for he recognizes that herein lies the true fascination of the subject'.[4] Certainly this would have been part of the delight of home construction; motivation could also be found in the opportunity to produce equipment better than that commonly on the market – possibly out of reach by reason of its cost.

The cost advantage was very markedly in favour of the home constructor, never so much so as when he was prepared to wind his own coils, cut pieces of tinfoil, waxed paper, etc., for condensers. For those who were unable, or unwilling, to go to such lengths, there was a large variety of well-finished parts on the market; in most cases there was no need for soldering as screw-terminals were commonly fitted; no need of metal work, virtually all sets were assembled on a wooden baseboard. Again, for those of an experimental frame of mind, the accumulation of a 'pool' of parts enabled more ambitious designs to be attempted, perhaps trying out new circuits available in legions of publications – complete books of circuits existed, e.g. 'Practical Valve Circuits' by John Scott-Taggart[5] – or on extending an existing set. Peto-Scott in particular facilitated this; all their components could be baseboard mounted to allow for endless permutations.[6] To take some of the tedium out of wiring up, other manufacturers, e.g. Metro-Vick, produced ready-wired 'unit' parts, (Radiobrix) e.g. tuner, valve amplifier, detector, etc., on individual identical boxes with terminals suitably arranged for easy inter-connection.[7] Such units still allowed for considerable experimental scope.

Many set manufacturers of the time went one better and sold kits of unassembled parts for some of their standard equipment complete with cabinet and the inevitable insulating panel. This last was almost invariably of ebonite and it was a great convenience to have this easily-blemished material already cut to size, drilled and even inscribed. Firms who produced such kits included McMichael,[8] Burndept[9] and Lamplugh.[10] A step further was to have complete sets blocked together on a major-unit system; with a crystal or 1 valve receiver; an H.F. amplifier with all its tuning: a power amplifier with all its couplings; each in a similar cabinet with terminals arranged for easy interconnection. These allowed only a little 'listener-participation' in the assembling of a larger receiver, but they were useful as the high cost of multi-valve sets put these beyond the reach of most potential purchasers; on the unit system, a very considerable receiver could be built up as finances allowed. It is an interesting comment on the perceived rate of circuit-development that such a system could be thought viable in the long term.

This great spectrum in the 'home construction' field makes it especially difficult to analyse the trends in this interesting aspect of wireless; much later Wireless World was having similar difficulties in arriving at a satisfactory definition; in an editorial it poses the question: 'What is a Kit Set?' 'It may sound absurd to ask such a question when wireless kit sets have been with us long enough for the name to have become almost a household expression. . . . manufacturers . . . have quite divergent views as to what they should comprise, with the result that purchasers are often surprised . . . to find that there are quite a lot of additional things to buy, the cost of which may amount to more than the kit set originally'. It comes to the decision that a kit '. . . should include every detail required, except

xxvi   THE WIRELESS WORLD AND RADIO REVIEW   OCTOBER 10, 1923

Build your own Set

The Ethophone Home Constructor No III

MANY people receive much more satisfaction in building their own wireless receiver than in purchasing one already made. Here is an opportunity for every one to build their own 3-valve set out of British guaranteed components. The set can be assembled quickly with a screwdriver, soldering iron and pliers only. When completed it has the appearance and performance of a factory made product of Burndept standard. The set will receive British Broadcast anywhere in the country. American reception is quite usual, but cannot be guaranteed. Continental Broadcast is almost a certainty. Will operate a loud speaker up to 80 miles from a Broadcast centre under normal conditions, but much greater distances have been obtained.

Write for descriptive pamphlet No. 234.

No. 532. Complete set of components from which to build a 3-valve receiver as illustrated (1 stage Radio Frequency, detector and note magnifier). Inclusive of Marconi Licence. **Price £12 0 0**

BRANCHES

LEEDS: The London Assurance House, Bond Place
NEWCASTLE-ON-TYNE: 7 St. Andrew's Buildings West. Gallowgate.
CARDIFF: 67 Queen Street.

HEAD DISTRIBUTING SERVICE DEPOTS.

ENGLAND.

BIRMINGHAM: C. S. Baynton, 133 New Street.
MANCHESTER: W. C. Barraclough, 61 Bridge Street.
BRIGHTON: H. J. Galliers, 32 St. James's Street
NOTTINGHAM: Pearson Bros., 54 Long Row.
YEOVIL: Western Counties Electrical & Engineering Co., Electricity House, Princess Street
BRISTOL: King & Co. Western Electric Works Park Row

SCOTLAND.

GLASGOW: W. A. C. Smith, Ltd. 93 Holm Street 246 Argyle Street.

IRELAND.

BELFAST: R. & S. Scott, Kingscourt, Wellington Place
DUBLIN: Dixon & Hempenstall, 17 Suffolk St.

CANADIAN OFFICES.

172 King Street West, Toronto

SERVICE DEPOTS IN ALL LOCALITIES

BURNDEPT LTD.
Head Office : Aldine House, Bedford Street. Strand. W.C.2.

BURNDEPT
WIRELESS APPARATUS

batteries and loudspeaker, to complete the installation',[11] possibly valves and cabinet could be excluded, if this were to be clearly stated.

Bearing in mind all the uncertainties, we shall survey the 'Home Constructor' scene over the decade, 1924–34. By 1924, with broadcasting well established and commercial apparatus readily available, home construction might be expected to be on the decline. That this was not so is evidenced by the large, and growing, popularity of journals, monthly, weekly – even daily paper – publishing new circuits, new designs, and catering frenetically for the needs of the amateur experimenter at all stages of expertise. Most towns of any size had 'wireless dealers', often cycle-shops or ironmongers, from whom components could be bought – often materials to make components. There was also a country-wide network of wireless clubs to interest and instruct all from the uninitiated novice to the hoary expert. Reliable figures reflecting the scale of all this activity are hard to come by, but some inkling may be obtained from the observation that during 1924, 120 wireless clubs more or less regularly reported their activities to Wireless World. These reports give an interesting glimpse of the wide scope of their interests and industry. In the same year, 158 different suppliers of components for amateur use found it worthwhile to advertise regularly in Wireless World, thus giving an insight into the value of the market. An interesting feature is that ex-Government materials were still available, even ex-R.A.F. 'C' type valves at 6/6 could be found,[12] and German types at 3/9.[13] These compared with new 'Radions' at 10/-; bright-emitters of all types, 12/6; dull-emitters 21/-; 60m/A filaments 30/-. The newer types differed from the old-stagers only in the filament current, so those on a limited budget could make considerable savings. This diversity of prices indicates why it was most common for receivers to be sold without valves; for a 4-valve set, savings of up to £4 could be made. Components of all kinds, both new and ex-government, were advertised with a wide diversity of prices. One of the chief reasons for home construction was undoubtedly economic; typical costs have already been discussed, sufficient to repeat here that a commercial crystal set, completely operational, was about £3.7.6; constructed to a published design it would cost about £1.12.6; for a four-valve set the figures were £44 and £10. Such differences must have been a powerful incentive to home construction.

Very few genuine kit sets were available at this time; to give some example of those featured we might choose the Peto-Scott version of the 2-valve + crystal reflex 'S.T. 100' (one of Scott-Taggart's published circuits). This was available in a flat-box form with engraved ebonite panel and all necessary components for £4.17.6 + 7/6 for the cabinet. The Marconi Royalty was paid and the components are noteworthy in that the variable condensers themselves were unassembled. The complete instrument 'aerial-tested' cost £8.5.0,[14] so the saving was £3. To these sums all 'accessories' would have to be added; valves, batteries, headphones, aerial and earth equipment etc. The same firm also had the 'Popular Wireless' reflex set, a 1-valve + crystal circuit typical of those displayed in magazines of the time. This retailed at £4.15.0. as a kit or £7.12.6. assembled, again a saving of about one third. By comparison the Sterling

'Anodion' 2-valve set cost £9.9.0 for the set only,[15] and the Ediswan 'Toovee' was £18, excluding valves.[16] This last, however, was housed in a very fine cabinet. The lowest prices for manufactured equipment were thus considerably in excess of those for kits.

Costs over the next few years declined slowly and the introduction of 0.1 amp filament valves between late 1925 and early 1926, with separate types for H.F., Det. and L.F., enabled a better performance to be obtained, although the improvements were not great. By 1926 'The country is spending some £15,000,000 a year on wireless, . . .'[17] and most listeners still used crystal sets. Since the licence-curve indicates a growth of some 450,000 listeners in 1926, this shows that the average sum paid for a new set and accessories must have been well under £33, when allowance is made for the sales of components. Crystal set prices had declined quite sharply; the 'Fellocryst Super' (Fellows Magneto Co), was only £1.15.0,[18] making a similar set from advertised parts costs about £1.7.6, so the balance of advantage is still with the constructor, but only just. The 4-valve McMichael was now £35.0.0 with no accessories, but similar sets (G.E.C.) could be had for about the same figure and complete; a similar design home-constructed could be made for about £15, a truly massive saving. The comparison is not quite fair to the commercial set, however, since with the new valves a 3-valve set would be as effective as the earlier 4-valve version; an A.J.S. 3-valve apparatus was about £20; home-made equivalent would be about £12. The financial benefits of home construction were thus declining; at the same time, higher efficiency valves required much more care in design and construction to achieve a satisfactory result.

For those intent on set-making at home, Wireless World gave two 'Buyer's Guides' – for L.F. coupling components;[19] for condensers, fixed or variable.[20] The extraordinary richness of the scene is shown by the numbers of manufacturers involved; 47 for L.F. transformers, 15 for L.F. chokes, 17 for anode resistances, 33 for fixed condensers and no less than 66 for variable condensers.

During late 1925 and over the next few years, reception conditions deteriorated rapidly, as more and yet more higher power continental stations joined the chaos. The problem was compounded in populous areas by increasing electrical interference. Selectivity became of paramount importance. In commercial receivers this led to a brief, if unhappy, flirtation with superhets, 5 exhibited at the Royal Albert Hall 1925,[21] perhaps a dozen at Olympia 1926;[22] numbers rather dwindled after this. More popular were sets with triode H.F. amplifiers, whether neutralized or not, in order to attain higher selectivity. Some idea of the effects of these changes may be judged from a passage from the 'Guide to the Show', '. . . Although these words are penned by one who derives much pleasure from the making-up of receiving sets, feeling perhaps that by doing so he is producing something better than the market offers, he is forced to realize by the models shown at this year's Exhibition that the status of the manufactured set has rapidly gained ground, and that the component market can only hold its own by creating new parts of outstanding merit. Yet the home constructor will always exist . . .'.[23] In order to cater for the

iv. ADVERTISEMENTS. THE WIRELESS WORLD AND RADIO REVIEW DECEMBER 1ST, 1926.

With the Bowyer-Lowe Super Heterodyne Kit, all the components are of the highest quality, and each carries an unqualified guarantee in keeping with Bowyer-Lowe Standards.

Built according to the diagram supplied, these components will make a set that will bring Radio to its peak of power and performance.

Include the Bowyer-Lowe Super Heterodyne Kit in your Gifts this Xmas—it will bring lasting and constant pleasure.

Supplied in an attractive Box.

£10.

List of Components supplied in the Bowyer-Lowe Super Heterodyne Kit :

1 Panel, $24 \times 8 \times \frac{1}{4}$, drilled, polished and engraved.
1 Panel, $5 \times 3 \times \frac{1}{4}$, drilled, polished and engraved.
1 Panel, $2 \times 2 \times \frac{1}{4}$, drilled, polished and engraved.
1 Panel, $22 \times 2\frac{1}{4} \times \frac{1}{4}$, drilled and polished.
1 Baseboard, $24 \times 9 \times \frac{1}{2}$, and four supports, $2\frac{1}{4} \times 2\frac{1}{4} \times \frac{3}{4}$.
1 Set of Four Matched Intermediate Frequency Transformers
1 Oscillator Coupler, 500–2,000 metres.
1 Base to hold above.
2 Square Law Condensers '0005.
2 Vernier Condensers.
1 Single Filament Control Jack.
7 Anticapacity Valve Holders
2 Brass Brackets
36 Tinned Copper Soldering Lugs.
24 Lengths 1/16 Square Tinned Copper Wire.

THE BOWYER-LOWE
ALL BRITISH SEVEN VALVE
SUPER-HETERODYNE KIT

THE "POPULAR" CONDENSER
is an example of Bowyer-Lowe precision and quality. Tested and guaranteed accurate before dispatch. Recommended by foremost experimenters. Supplied with 3 in. dial.
'0003 M.F. .. **10**/-
'0005 M.F. .. **10**/6

THE SUPER HETERODYNE KIT

THE BOWYER-LOWE RADIO NEWS
contains particulars and illustrations of all our trustworthy components — also two constructional articles of interest to amateurs.
A portable set and a four valve receiver are fully illustrated and described.
$1\frac{1}{2}$d. in stamps will secure your copy — Get it now.

Bowyer-Lowe
TESTED RADIO APPARATUS

Announcement by the Bowyer-Lowe Co., Ltd., Letchworth, Herts.

AUGUST 31ST, 1927.   THE WIRELESS WORLD   ADVERTISEMENTS.   3

A Satisfactory Home-Built Set

Satisfactory to build
Satisfactory to hear
Satisfactory in cost

The illustrations show an excellent and easily constructed set, for three valves and resistance capacity coupling built entirely of

Cosmos
Components & Valves

Prices are given for the whole range of types and sizes of each of the components used. ASK YOUR DEALER FOR A COPY OF THE CONSTRUCTION LEAFLET, No. 7117 I. COMPLETE WITH DRILLING TEMPLATE. **Note Reduced Prices.**

"COSMOS" LOW LOSS SQUARE LAW CONDENSER

SLOW MOTION	·00025 MFD	10/6
	·0005 ,,	11/6
ORDINARY	·00025 MFD	8/6
	·0005 ,,	7/6

"COSMOS" RESISTANCE COUPLING UNIT

TYPE "O" 8/6 (UNIT ALONE)
TYPE "V" 10/6 (WITH VALVE HOLDER)

"COSMOS" SPRING VALVE HOLDER ALONE 2/9 EACH

"COSMOS" PERMACON EACH

·0001, ·0002, ·0005 MFD	1/6
·0003 (WITH CLIPS) & ·001 MFD	1/8
·002 MFD	1/10
·005 ,,	2/8
·01 ,,	3/9

"COSMOS" 2-COIL HOLDER

PANEL MOUNT'G 1/3
WITH BASE (AS ILLUSTRATED) 2/-

The "COSMOS" RHEOSTAT

SINGLE WOUND 6·0 OHMS 1·0 AMP 4/6
DOUBLE WOUND 20 OHMS ·4 AMPS OR 34 OHMS ·2 AMPS 5/-
POTENTIOMETER 300 OHMS 6/-

Advt. of METRO-VICK SUPPLIES LIMITED - LONDON

changing needs of the home-constructor, some firms produced massive new kit sets. One Marconiphone superhet, illustrated in Wireless World was a 7-valve monster erected on a baseboard.[24] Much more than ordinary skill would have been needed to make a success of such a venture. The same firm also gave details of several simpler receivers. None of these were really kit sets in that the constructor simply bought the specified components and assembled them, if he could, following the circuit and instructions in the accompanying booklet; a far cry from just a few years previously when a cardboard tube, a few ounces of 20 S.W.G. d.c.c. wire, a crystal and a set of head telephones were all the essentials! Igranic produced complete kits, of American origin, along similar lines.

Other kit sets on a less grandiose scale were available; Blackadda 'Radio Building System' 'Install the system that is readily converted to any new circuit . . . parts can be used over and over again . . . even the wiring is supplied in six standardized lengths . . . construction is accomplished by the aid of a small box spanner . . . construction by numbers'.[25] This appears to have been a form of Radio Meccano. Ediswan had 'The R.C. Threesome' 'If you can use a screwdriver you can make this set in one evening . . . the set can be made for £3, or less'.[26] On a grander scale in the Bowyer-Lowe 'All British seven valve Super-Heterodyne Kit'. The kit is illustrated 'Supplied in an attractive Box. £10'.[27] The list of contents includes lengths of square wire – a relic of an earlier age; the whole thing would have made a quite remarkable Christmas gift – as suggested. In view of the scale of these sets, it is no wonder the writer of the 'Review' article was sounding rather defeated!

The march of progress continued unabated; Mullard now entered the race 'Set Construction for the Novice. "Radio for the Million" is the title of an interesting quarterly publication, the first number of which has just been issued by The Mullard Wireless Service Co., Ltd'. 'The aim of the new periodical is to provide those who have no technical knowledge of wireless with simple details for building broadcast receivers',[28] – and presumably sell Mullard valves. Four sets are mentioned and blueprints are included. Reference to Vol. 1 No. 2 appears three months later[29] and others occur at intervals thereafter. This seems to be the first foray of a valve maker into the set market. Right through the year various kits, part-kits, circuits etc., were advertised; a selected few will serve to present the scene:

Wireless World; a review of the 'Ediswan R.C. Threesome'.[30] It was designed around R/C coupling using two high impedance triodes and an output valve; total cost about £9. The Cosmos 3 valve set. This appears to have two forms: one for home constructors and mounted on a wooden baseboard;[31] a commercially-produced one making use of a moulded vulcanite base.[32] Wireless World; 'Owing to the success of the "M.H." five-valve Supersonic Block Unit, Messrs. L. McMichael Ltd., . . . have issued constructional details for a six – or seven – valve portable';[33] cost, complete, was £21.16.0. This appears to have been an attempt to exploit a unit developed for a commercial set and increase sales.

For the 1927 Show both I.H. and S.G. valves were available; the situation for the home-constructor must have seemed as if it were slipping out of

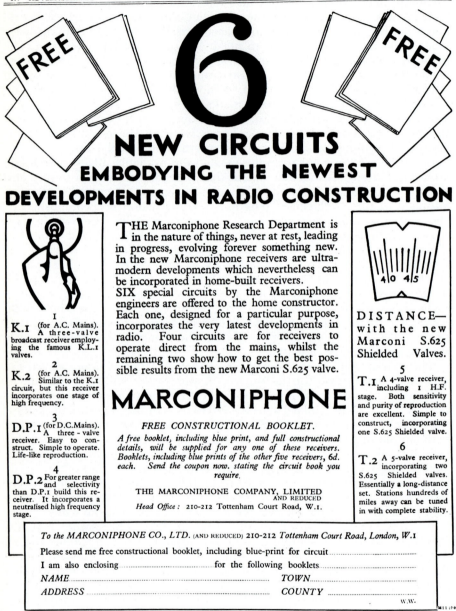

control. Marconiphone continued to try to help the addict and improve sales by providing blueprints and free constructional booklets of circuits for mains receivers incorporating their I.H. 'K' valves, and battery outfits using the new S625 S.G. valve. Just how realistic it was to expect the average constructor to cope with the hitherto largely unexplored fields of mains operated apparatus, extensive screening, etc., is very hard to assess, certainly one of these sets, illustrated in Wireless World appears very complex.[34] Other manufacturers, about 7 in all, showed kits of one form or another at the Show.

An entirely different approach was made by Cossor who introduced what might be regarded as the prototype kit set (sold complete from 1928) as produced by a major manufacturer, the 'Melody Maker'. This is described as a det. 2 L.F. set.[35] Cossor valves are of course used, but the other components came from other manufacturers. The Cossor advertisement claims 'Radio Revolutionized – not a booklet – not a Blueprint BUT a wonderful new system of set building. The makers of Cossor Valves have evolved an entirely new system of Set building which ensures perfect results without skill. Soldering is abolished, technicalities have been eliminated. Success is certain. . . . you will quickly want to discard your out-of-date Receiver'.[36] A later advertisement continues the story 'The new B.B.C. Alternative programmes demand a new standard in Receiver performance. Only the Receiving Set which is "Razor-sharp" in tuning will be sufficiently selective to tune out the unwanted local station in favour of the one which is required. . . . The wonderful new Cossor "Melody Maker" is . . . highly selective . . . It is a real "alternative programme" Receiver, for if its owner is dissatisfied with the B.B.C. programmes a large number of Continental Stations . . . are always available at full loudspeaker strength'.[37] These are remarkable claims for a receiver with no H.F. stage and only one tuned circuit.

The McMichael Company quickly took up the challenge with a complete kit for their 'Home Assembly "Dimic Three"', advertised as 'It's an Easy Step – from Workbench to Drawing Room!'.[38] The price was £11 + £1.17.6 Royalties. Mullard were not to be left out; an advertisement for their kit set, carried the legend 'Mullard Master 3' which could be built in one hour! 'Without any knowledge, and using only a small screwdriver, you can build . . . the Mullard Master Three . . . a complete set of wires cut to size. . . . put them under the terminals and screw down tight. . . . no soldering needed'.[39] Like the Cossor set, this Mullard kit was for a rather primitive and outmoded receiver, hardly adequate for the prevailing conditions and certainly not fitted for the future. These two sets are also of considerable significance as they showed two valve manufacturers testing the market for receivers; Cossor went on to become large-scale makers, Mullard to a lesser degree. Whatever their limitations, these sets, especially the Melody Maker, seem to have captured the public imagination.

The early part of 1928 was apparently very quiet for kit sets and possibly for home construction too; perhaps the rash of new valves, new circuits and new techniques overwhelmed potential constructors. This view is supported by a Wireless World editorial 'From the point of view of the

IN ONE HOUR
Wonderful Radio
for every Home

Mullard

The
Master
Three

JUST THINK OF IT! Without any knowledge, and using only a small screw driver, you can build for yourself the Mullard Master Three—the finest three-valve radio receiver yet designed.

EASY TO BUILD. You cannot make a mistake —the Plan of Assembly supplied free is drawn to full size and all you have to do is to mark through the position of each component on to the baseboard. The packet of "Master Three" A.B.C. Connecting Links contains a complete set of wires cut to size. You have only to put them under the terminals and screw down tight. **THERE IS NO SOLDERING NEEDED.**

EASY TO HANDLE. There are only two dials, one to bring in stations and the other to bring them up to strength. Wherever you may live you can bring in 6 or more broadcast programmes as easily as setting the hands of your watch.

EASY TO BUY. Your nearest radio dealer stocks the complete list of components required. Tell him you want to build the Mullard Master-Three—he'll know what you want.

Build the Mullard Master Three now!

The Editor, "Radio for the Million."
63, Lincoln's Inn Fields, London, W.C.2
Please send me FREE complete in-
structions and Simplified Plan of
Assembly for the MULLARD MASTER
THREE with No. 5
"RADIO FOR THE MILLION."

NAME (Block Letters) ....

ADDRESS ....
....   i WW

Mullard
MASTER · RADIO

22 ADVERTISEMENTS   THE WIRELESS WORLD   OCTOBER 31ST, 1928.

BUYS ALL THE PARTS

£7·15·0

for the wonderful NEW

Cossor 'Melody Maker'

FOR £7. 15. 0. you **can** buy all the parts for the wonderful new Cossor Melody Maker. They are sold in a sealed box—sealed to prevent substitution of inferior **or** unsuitable components—sealed to ensure your obtaining the right parts and those only. Everything is included—the handsome all-metal cabinet, the three Cossor Valves, the wire and even the simple tools. There is nothing more to buy. Get to know all about this wonderful new Cossor Melody Maker.

← **Fill in this coupon NOW!**

Please send me free of charge one of your Constructor Envelopes which tells me how to build the new Cossor "Melody Maker."

Name..................................

Address..................................

W.W. 31 10 28

OCTOBER 31ST, 1928.   THE WIRELESS WORLD   ADVERTISEMENTS. 1

ONE DIAL TUNING

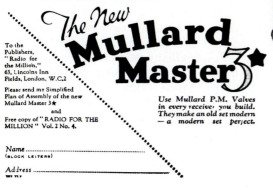

amateur . . . it appeared that, early in the present year, the field of radio development had been exploited to the full. . . . Contrary to this point of view, new and far-reaching developments . . . have been created, . . . the enthusiast is almost faced with difficulty in keeping up to date'. It then goes on to list a series of ways in which an amateur needs to be able to apply his knowledge and understanding to interpreting valve characteristics etc. The new valves, it is claimed, will result in '. . . drastic modification of our present circuits, . . .'.[40] There also seemed to be a downturn in wireless society activity; the annual meeting of the Bradford Radio Society on 7 February 1928 was given over to a discussion on how to maintain enthusiasm through the coming year.[41] Other societies seem to be involved in social rather than scientific events, still others turned to transmitting work.

The distinct feeling from the evidence accumulated is that wireless was beginning to move into a new age where a much greater technical knowledge was necessary unless recourse was to be had to merely screwing together boxed kits. And indeed kit sets appeared. At the Olympia Show Cossor brought out an updated 'Melody Maker'. The description of the set shows that one of the new S.G. valves was incorporated and the set was advertised as 'Simple as Meccano . . . Built in 90 minutes'.[42&43] The cost was £7.15.0[44] and the set, although still very simple, was a little more in keeping with the ever more difficult reception conditions. G.E.C. ripostes immediately with its curious ready-built Victor 3 under the headline 'Don''t trouble to make your own set when you can buy this famous Geophone for less-'[45] – it cost just £6.17.6, but had no H.F. stage. The 'Mullard Master 3*' was now advertised as having 'One Dial Tuning',[46] also the 'Osram Music Magnet'[47] in an S.G.-det.L.F. form made its bow, priced at £8.11.3 and with more advanced circuitry than the 'Melody Maker'. Others followed, thick and fast: Formo S.G. 3;[48] Peerless 'Resonic 2', £3.11.0;[49] 'Eddystone' S.G. H.F. Short wave kit, £8.10.0.[50]

Comments in the 'Trends of Progress'[51] review illustrate how commercial set design was moving from the familiar baseboard and going to metal chassis; there was elaborate metal screening for H.F. stages, more – and more complex – decoupling; all unfamiliar features likely to further discourage ordinary home constructors and encourage kits; and they came. Wireless World over the 1929–30 period ran a series of regular features 'Kit Constructor's Notes' in which various kit sets were described, discussed and analysed. There were also separate sections giving advice on correction of problems experienced by readers. A list of the topics is given below:

Wireless World Kit Constructor's Notes

23 Jan 1929	p. 92	'The New Cossor Melody Maker'
20 Feb 1929	p. 202	'Osram Music Magnet'
27 Feb 1929	p. 221	Cossor Melody Maker on Ultra Short Wavelengths
20 Mar 1929	p. 306	'Mullard Master 3*'
3 Apr 1929	p. 361	Six-Sixty 'Mystery Receiver'
24 Apr 1929	p. 433	Ferranti 'Screened Grid Three'

1 May 1929	p. 463	Formo 'Screened Grid Three'
22 May 1929	p. 532	Dubilier 'Toreador Screened Grid Four'
19 Jun 1929	p. 641	McMichael 'Home Constructor's Screened Three'
10 Jul 1929	p. 38	The Mullard 'S.G.P. Master Three'
24 Jul 1929	p. 83	Valve Filament Risks and Some Safety Precautions
31 Jul 1929	p. 103	Kit Set Circuit Diagrams and How to Understand Them.
18 Sep 1929	p. 260	'Wireless World Kit' Set
9 Oct 1929	p. 403	The New Kit Sets at Olympia
27 Nov 1929	p. 582	'Wireless World' A.C. Mains Kit Set
18 Dec 1929	p. 674	'New Osram Music Magnet'
23 Apr 1930	p. 438	'Lewcos' 3 Valve Kit Assembly
25 Jun 1930	p. 674	'Lotus' all mains adaptor for 'Music Magnet'
13 Aug 1930	p. 149	Next Season's 'Osram Music Magnet'
15 Oct 1930	p. 454	Osram Music Magnet Four

Ten manufacturers are represented, some with several different versions. An article entitled 'The New Kit Sets' taken from those on show at Olympia 1929 lists in addition; Brown (2 battery, 2 mains); Bullphone (2 battery); Cossor (pre-wired H.F. stage, A.C. version with pre-wired mains unit); Ferranti (S.G.3, S.G.4); Formo (S.G.3); Lewcos (Chassis rather than kit, battery and mains); Lissen (S.G.3, battery and mains, also S.W.); Lotus (S.G.P. battery and A.C.); Mullard (Orgola 3, S.G.; A.C. versions, Orgola Senior 6-valve); Osram (New Music Magnet); Varley (R.D.3, det. 2 L.F.). There was thus a wealth of opportunity and the kit set was clearly flourishing.

The Wireless World article on 'The New Cossor Melody Maker' makes a number of interesting points 'A Wireless Rip van Winkle, oblivious to developments in the amateur field since, say 1922, would . . . probably be even more intrigued by the way in which the set builder's difficulties have been removed than by the technical progress . . .'. The article continues 'Kit construction is not new, but what changes it has undergone!', it refers to the period just before broadcasting when a 'bargain offer' of a set of parts for making a tuner cost £5 'Nowadays we expect to get all necessary components for a simple receiver for something less than that, . . .'. 'The new Cossor Melody Maker . . . a kit of parts that can be assembled and wired by the veriest novice. Everything, excepting batteries and loudspeaker, is included and there is no need for . . . drilling or sawing; . . . terminals are used . . . in place of soldered connections. Even the boring of holes for wood screws is avoided . . .' 'Needless to say, the receiver can be wired without any knowledge of theoretical diagrams, . . .'.[52] The set was, in fact, very simple, using plug-in coils with tapped winding as in olden days, separate tuning capacitors, and relying on reaction to give some sensitivity and selectivity. A design fault (unearthed screen) and certain lack of selectivity was commented upon, but the kit was cheap, £7.15.0. By scanning through the pages of Wireless World it was possible to estimate the cost of buying all the parts; a figure of about £6 was reached for components, valves, etc., leaving under £2 for cabinet, baseboard, screws, wire etc. The economics of producing such a set must have been very marginal; even the ready-built mass-produced and much simpler G.E.C. Victor 3 was £6.17.6, and as G.E.C. pointed out, why bother with a kit set anyway! The 'Music Magnet' was a distinctly more advanced design with 'ganged' single dial tuning, wavechange switch and a substantially better performance, yet it cost only £8.11.3. The 'Mullard Master 3*' was, if anything, even more primitive than the Melody Maker. All these sets were very similar; they did allow a complete novice to produce a working receiver, but the performance, with the honourable exception of the Music Magnet, must have been quite poor and the cost saving for the constructor was at best marginal. For the larger manufacturers, it must be doubted if they made any significant profit at all; perhaps kits were more in the nature of an advertisement, bringing their name before the public; in that they were undoubtedly successful. Cossor, in particular, moved on to become manufacturers on a very great and integrated scale.

One or two of the kits from the smaller firms had special features which were probably designed to appeal to a particular market. About 21 in all were advertised in Wireless World for 1929. Amongst these was the Lewcos set which was almost a complete receiver ready-made; it needed only a curious collection of external parts added to complete it, three variable capacitors and an on/off switch! The idea seems to have been that the purchaser could update an old receiver, probably housed in a substantial piece of furniture, by removing the bulk of the 'works' and substituting the complete unit. Most early sets would have at least two tuning capacitors and a reaction control attached to the front panel, it would be a rare set indeed that did not have an on/off switch, so these external controls would usually be available. As we have seen already, many of the earlier sets were costly and a substantial part of that cost lay in the expense of the 'furniture' in which they were housed. This concept then enabled the familiar furniture to be retained and modernized. It seems to have had a short life; partly due, most likely, to the very narrow range of potential purchasers. The prospects of disembowelling an existing set and rewiring a new unit into it cannot have been within the capabilities of many of those who had initially bought an expensive piece of apparatus. The Ready Radio 'Empire Link' was also a curious kit, intended to cover S.W. as well as broadcast bands. Cash price was £11.11.0, but a most odd feature was the offer to buy back 'your old Receiver – in part exchange'. Ready Radio were stated to be the 'Sole Distributors', but no hint as to the origin of the kit is given. Another oddity was by 'Six-Sixty', 'Convert your set to A.C. Mains'.[53] By means of ingenious adaptors, an existing battery set could be partially rewired to accept Six-Sixty A.C. valves and a power unit was supplied at a total cost of £8.5.0.

The one exception to the feeling that these kit sets and adaptations were in general a rather unsatisfactory lot must be the Osram 'New Music Magnet Four' of 1930. The Coventry factory where this was produced is illustrated in Wireless World[54] and the preview of the set includes the phrase '. . . the manufacturers state that the conventional H.F.-det.-L.F. set is insufficiently selective for present day conditions . . .'.[55] This was indeed

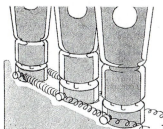

true, as we have seen before. The updated version had two S.G. H.F. stages, waveband switching and 'ganged' tuning for all three tuned circuits, a feature not common even in factory-made commercial receivers of the time. The review was very favourable including such phrases as 'Selectivity is . . . more than sufficient for ordinary conditions'. 'The "ganged" tuning system works very satisfactorily, . . .'. '. . . the chassis and details appertaining to it are beyond criticism . . .'. The complete kit cost £11.15.0, drawing the comment 'This price seems to be remarkably low, . . .'.[56] It certainly was! Unless production runs were very large indeed, it is difficult to see how any profit could have been made on this set.

One feature, just touched on so far, is that a number of these kits, complete and incomplete, were designed to be used on A.C. mains valves. This could well have been an advantage in boosting sales and certainly of benefit to home constructors as many looked askance at the terrors, real or imaginary, of mains equipment. Wireless World made several points 'In a few years" time receivers operated from A.C. mains will be practically universal, . . .'. After discussing American experience and the general convenience of mains operated equipment, the writer (W. T. Cocking) plays down the difficulties 'Many people seem to think that a mains-operated set is difficult to design and build. . . . but the difficulties are so slight that they need not deter anyone capable of designing a battery set from attempting a mains set.'.[57] The operative word here is surely 'designing'; from the evidence of the 'Query' pages of all the magazines of the time, many of the enquirers were far from being in a position to 'design' any set. A word of caution about mains operation was reiterated by the B.B.C. 'The B.B.C. is anxious to dissuade listeners from themselves attempting to construct a mains unit, unless they are used to the handling of power mains and are thoroughly conversant with the dangers and difficulty which it implies':[58] perhaps a more realistic view. Cossor seemed to think so, too; their A.C. mains version of the 'Melody Maker' came with the mains unit completely wired and enclosed.

For those not content with a kit construction there was great scope. 'Modern Wireless' asserted '. . . the home constructor is catered for much better at Olympia this year . . .'. It also gives a word of cheer to 'proper' home constructors from Capt. H. J. Round who apparently announced '. . . at least twenty-five percent efficiency goes West in production . . .' and continued '. . . if the home constructor did not gain at least that twenty-five percent – and gain, too, in always being a bit ahead in the way of circuits – home construction would not be as popular as it is, for prices are down this year to what must be rock bottom levels'.[59] This last is a reference to commercial sets shown at Olympia for which, indeed, prices had fallen quite dramatically, as we have seen. To support the eager army of 'genuine' home constructors, there was an enormous supply-dump of material, no fewer than 148 component firms advertised in Wireless World through 1930.

By the start of 1931, after the richness of 1930, the kit scene seems to have been very quiet. Of sets from the major manufacturers the 'New Music Magnet Four', was the only one to be advertised to any great extent, although Cossor, Mullard, Ferranti and a newcomer Voltron are occasionally mentioned. Perhaps the observation that at the 1930 show mains receivers outnumbered battery ones for the first time may have been a pointer in the wind; in spite of Cocking's assertions, many found the construction of a mains set a daunting prospect. Also, if a mains set had been in the household, all the tyranny of battery charging, so thankfully dispensed with, would be unlikely to be resumed. For home constructors in general a 'Show Forecast' article, makes an interesting statement which contains a warning 'The time when the annual Wireless Show was the occasion to reveal a glittering array of new components has long since passed'.[60] There was also a continuation of the decline of club activities; this had been apparent in 1930 when only 30 were mentioned in Wireless World, but it is emphasized in an editorial and in the correspondence columns early in 1931. The discussion was sparked off by a letter 'Radio Dealers and Radio Societies' by a Manchester reader, J. Baggs. He starts with the claim 'During the last few years the membership of the many radio societies which used to exist throughout the country has fallen to a very great extent, so that in many cases they are supported only by the really hard-boiled enthusiast. This change is undoubtedly due not to any diminution in the interest in radio, but to the fact that its technical development has now reached a stage where a certain amount of knowledge is necessary if progress is to be made by any amateur, and, in consequence, it tends to become rather beyond the understanding of the man who used to be the backbone of the average radio society'.[61]

The solution suggested was to form new societies incorporating radio traders as well as serious enthusiasts, thus permitting a higher technical standard of discussion and attracting lecturers from major manufacturers. The following week's editorial is devoted to the topic; it accepts that it is undoubtedly true that membership of many formerly active societies has fallen to a great extent, but there are 'outstanding exceptions' and suggests that the obvious decline is due more to the policy pursued by the existing societies. Broadly, the editorial supports the view presented by the correspondent. One Robert Reid Jones of Liverpool added his contribution, '. . . I think the reason . . . is due to the introduction . . . of the "Kit set". The 20 wires and no solder ideas has taken the interest away, . . .'. The following letter, by G. W. Heath, Assistant Hon. Sec. of the Hackney Radio and Physical Society, agreed '. . . the membership of the societies has dropped to the "really hard-boiled enthusiasts" . . .' but suggested 'All the people who wanted to know how to wind a variometer have been shown – and who wants to, anyway, in these days of cheap components and blueprints?'.[62] The topic was continued in a letter by the President of the R.S.G.B., H. Bevan Swift, in which he confirmed the decline 'One of the saddest features in British amateur radio history is the demise of so many of the early societies . . .'. He saw one of the causes of apathy being in the membership. 'These must not be men to whom interest in radio is but a mere passing fancy. . . . generally found in the tyro who, having succeeded in building himself a radio set, probably from a set of kit parts, . . . promptly poses as an authority . . .'.[63]

So there were five different appraisals all agreeing on the decline in membership and enthusiasm, but advancing different theories for the reasons. It is probable that all contained an element of the truth in their analyses; technical progress taking the subject beyond the abilities of the man-in-the-street; kit sets removing the interest, giving adequate results but leaving the builder knowing little of the subject; cheap components taking away the excitement of making your own and finding it works. It is likely that such informed opinions reflected the situation in the country as a whole, indicating a potential collapse in the numbers of home constructors. It is perhaps interesting to note that Wireless World was at this time trying to reverse the tide by running regular 'New Readers' issues.

Not everyone was of the same depressing opinion; 'The Wireless Constructor' under 'Auto Electric-Devices Ltd', warns 'Pessimists who thought that the day of the home constructor was over should pay attention to this display (at the Olympia Exhibition) which shows how he flourishes and is still catered for with the utmost keenness';[64] and indeed revival of interest on the part of the manufacturers is apparent '. . . there were certainly more "kit" sets for home assembly than ever before, including complete sets of parts, boxed together and sold complete, and also a number of receiver circuits sponsored by the manufacturer producing the majority of components in them'.[65] This last phrase may just possibly indicate the reason for this second flowering of kits as so many of the set-makers now used mass production methods, no longer an assemblage of finely-finished components, so the component manufactuers had to seek new markets. As an example, four sets by the well-known Telsen component-manufacturing company are mentioned and they continued in this line for some time after the others had given up. The Osram 'Music Magnet Four' was available, also in an A.C. version, 'Radio for the Million' seems to have become divorced from Mullard, it was now accredited to United Radio Manufacturers Ltd., but Mullard was still offering a 'Mullard Three'. Ferranti had not a complete kit, only major parts of their own manufacture requiring additional components to be bought. Cossor was now offering the 'Empire Melody Maker'. The Wireless Constructor carries the paragraph 'It is interesting to note that the staggering figure of 350,000 Melody Makers have been issued to the public, and the latest three-valve S.G. set for £6.15.0, is certainly likely to increase that number'. Reference is also made to the A.C. mains versions 'your broadcast programmes will cost you less than 1d. a day, and the constructional charts make the job of assembly a sheer delight'.[66]

That a resurgence in kit sets was established is evident from Wireless World.[67] Many in fact had been updated for the 1932 Show and appeared with integral loudspeaker in a cabinet following the trend in commercial receivers of the time. The Music Magnet was there, now in a smart new bakelite case. The valve complement had been reduced to three, H.F.-det.-L.F., but the performance was good as the detector was now a S.G. valve. The price quoted in 'The North British Machine Co. Ltd.' catalogue for 1932, was £9.9.0. The older 4-valve version was also there, now at £10.4.0., and the full A.C. form is £15.11.0. Cossor Melody Maker was present, now in four models ranging from £6.7.6. for 3-valves without speaker up to £11.15.0 for an A.C. apparatus complete in the cabinet with the loudspeaker. This price was considerably less than comparable ready-made sets which appeared to sell at around 14 guineas. Lissen had joined the ranks and Ferranti had several models.

The following year, 1933, showed the resurgence faltering. Although at the beginning of the year The Wireless Constructor could claim 'At the time of writing this editorial, over 210,000 copies of The Wireless Constructor containing details of Mr Scott-Taggart's "ST 400" have been sold, and still the demand goes on',[68] by Show time Wireless World noted '. . . there are fewer receivers available this year as kits of parts, . . .'. Even in the reduced numbers, standardization was appearing 'Cossor Melody Makers are produced in several models, . . . designed for mounting on the same steel chassis'.[69] There were exceptions to the decline; Lissen, possibly as a latecomer to the scene, had a comprehensive 'Skyscraper' range. A four-valve, Q.P.P. battery set covering 12m to 2000m retailed at only £5.12.6. They even had a 7-valve superhet.

Moving on to 1934, a quotation from the B.B.C. Year Book shows that by then a home-made receiver was '. . . a comparative rarity'. 'It is no longer able to compete successfully as regards either price or performance with the product of the manufacturer who has adopted the press tool and the drilling jig to produce the metal chassis and quantity production methods . . .'.[70] In kit sets the scene appeared similar; Wireless World in reviewing the Show for that year asserted that 'Kit sets for the home constructor are exhibited by Lissen and Cossor'.[71] In this bald statement is encapsulated the epitaph of the kit era. A glance at the Lissen set shows it essentially to have been a commercial set in unwired form;[72] Cossor appears to have gone the same route. In a curious way we go full circle, for Wireless World ran a series of articles aimed at the home constructor which echoed the activities at the beginning of this chapter, indeed back to the beginning of broadcasting. An editorial makes the statement 'In the early days the constructor had very few components available ready-made for his use, and consequently he had to set about making nearly every part himself. Many components of to-day are really too complicated to be within the scope of the ordinary constructor to build up, whilst others require such careful matching as to necessitate the use of special apparatus'.[73] An article asserts 'There are no particular difficulties in the home construction of certain simple components where quite ordinary tools will suffice.' 'It is advisable, however, to restrict activities to the making of those parts that can be built to a specification and then put into use without elaborate measurements and adjustments to ensure their electrical constants being correct'. Within this category are included: 'coils, preferably for use in sets fitted with separate tuning condensers, as the accurate matching of inductances for "gang" tuning is a tedious business unless a fair amount of apparatus is available. We might include, also, H.F. chokes, as these are comparatively easy to make, mains transformers, output transformers and L.F. chokes. Intervalve transformers might even be attempted, as their construction is well within the capabilities of the amateur, though the design should be correct. To these can be added short-wave parts'. Advice is given on suitable materials; 'Out-of date components stripped down will often provide much useful material which can be augmented where necessary by new parts, as coil screens, formers, transformer material, and the sundry small items needed are readily obtainable'.[74] Then followed a list of manufacturers of coil screens, formers and transformer laminations. This list of components deemed suitable for home construction was very restricted and it is to be doubted if the average hobbyist could realistically produce mains transformers. Most of the rest of the issue was concerned with a general survey of some of the commercially available apparatus; tuning coils, I.F. transformers, L.F. transformers and chokes, loud-speakers. Articles on soldering and simple component testing were included. The whole edition paints a rather sombre picture of the opportunities available for the home constructor; he seems to be con-strained to use in the main commercially available components specifically designed for use in particular, pre-determined, circuits. This trend was to continue with the production of coil-packs, even complete tuning units, condenser included, all ready wired, becoming common in the future.

In the foregoing, there is nothing of the scope for adventure and worthwhile experimentation so evident a decade earlier. The whole scene had changed; broadcasting stations were so numerous and so powerful that only genuine 'knife-edge' selectivity would do; carefully-matched tuning coils and 'ganged' condensers allowed for very little latitude in experi-mentation; good receiver design was largely conditioned by the highly efficient but very specialized valve-types available, virtually trouble – and maintenance – free commercial mains sets abounded; cabinet design and finish had reached a standard difficult to match with amateur construction; above all manufactured set prices were so low that little incentive remained. In only a few fields not yet fully catered for by the assiduous activities of the commercial manufacturers could an outlet for some individual talent still prove satisfying; a resurgence of interest in short-wave work was apparent and the evolution of television provided scope for new experi-mentation. In many cases interest switched and the 'hard boiled amateurs' transformed themselves into 'hams', concerned with amateur transmitting topics. This is not to say that interest and indeed development died overnight. Some 70 component suppliers advertised regularly in Wireless World during 1934; Telsen in particular developed a large-scale industry in kits and component supplies; some half-dozen kit sets were still being advertised; but the halcyon days were over and a rapid decline is evident. Even Practical and Amateur Wireless, always concerned with home construction topics, was admitting the trend by 1936. An article headed 'To Our Readers – and the Trade' upbraids the manufacturers for ignoring the needs of the home constructor. Although the article specifically talks about television it clearly applies to radio also and it claims '. . . this journal continues to be the leader in the field, that it has the largest net sales in that field, and that its considerable circulation . . . should indicate . . . that the construction market is by no means dead . . . very few genuine constructors have deserted our ranks . . .'.[75] 'Thermion' picks up the theme; 'The numbers of firms specifically catering for the home constructor has shrunk to meagre proportions'.[76] An article 'The Constructor and the Show' presents a different picture; 'From a careful survey of Radiolympia it is evident that manufacturers are catering for the constructor even better than before',[77] but the collapse of the local component supply network is clear from 'Thermion's' comments 'The dealer today does not seem to care two hoots about home construction; he wants the parts to be ordered and he will take about a fortnight to get them for you'.[78] And so, again we have come full circle back to the conditions pictured at the very start of the chapter.

It is interesting to conclude with reference to the article 'Four Years of Home Construction' in Practical and Amateur Wireless. The journal was four years old and regaled its readers with a list of 'firsts' which it claimed for itself over that period: plywood chassis construction – 'copied by others!'; metallized wood chassis material – 'undoubtedly here to stay.'; Class B and Q.P.P. output stages – 'almost a year later . . . there was any number of commercial receivers employing Class B.'; iron-core tuning coils – 'Here again, the constructor led, and set manufacturers followed.'; automatic volume control – 'One of the first highly sensitive and powerful receivers to employ A.V.C. was the "Luxus Superhet," . . . Thereafter . . . A.V.C. has been almost universally adopted by manufacturers . . .'; midget components – 'In this respect we claim to have been pioneers, . . .'[79] The article concludes by forecasting that 'television' and 'ultra-short' will figure prominently in the future. With this last conclusion, at least, we can agree!

As a final quotation which elegantly sums up part at least of the reasons for the eclipse of home construction, we may draw from an article by John Scott-Taggart grandly titled 'My Plans'. In connection with '. . . a man who gave up active interest in 1926, I asked him why he had not taken up his old pastime. He replied: "Oh, I've occasionally bought an odd copy of a wireless paper – but it only makes me realize what big changes have taken place: different words, different technique, different valves!" In fact, he was rather scared at screening, screened-grid valves, band-pass tuning, decoupling circuits, pentodes and other features'. Scott-Taggart aimed to regenerate his enthusiasm with a series of new, fresh and explanatory articles and 'convince the person who feels himself a bit "rusty" that wireless as a hobby is now cheaper, is easier than ever it was and has infinitely more to offer in results'.[80] We can probably judge the success of the great man[81] by the subsequent events.

We may say that the kit set had a long reign from primitive beginnings at – or even before – the start of broadcasting right through to 1939. There were apparently three phases: (1) the products of comparatively small firms, usually quite independent, right up till late 1927; (2) the Cossor Melody Maker introduced a wave of increasingly sophisticated kits by major manufacturers which persisted, with a hiatus in 1930, until 1934; (3) a decline with reversion mainly to smaller firms again, and with more emphasis on developments outside the normal broadcast band.

References

1   Wireless World, 10 October 1923, pp. 50ff
2   Wireless World, 21 July 1926, pp. 97ff
3   A. Hyatt Verrill; Harper, N.Y., 1922, pp. 102ff
4   Wireless World, 28 May 1924, editorial
5   Radio Press, 1923
6   Wireless World, 31 October 1923, p. xxiv
7   Wireless World, 7 November 1923, p. v
8   Wireless World, 3 October 1923, p. xxii
9   Wireless World, 10 October 1923, p. xxvi
10 Wireless World, 17 October 1923, p. xvii
11 Wireless World, 4 November 1932, editorial
12 Wireless World, 28 May 1924, p. xxiv
13 Wireless World, 16 April 1924, p. xxv
14 Wireless World, 28 May 1924, p. xxvii
15 Wireless World, 21 May 1924, p. xv
16 Wireless World, 28 May 1924, p. xxii
17 Wireless World, 17 March 1926, p. 414
18 Wireless World, 10 February 1926, p. 200
19 Wireless World, 26 May 1926, p. 709
20 Wireless World, 23 June 1926, p. 855
21 Wireless World, 16 September 1925, pp. 355ff
22 Wireless World, 15 September 1926, pp. 381ff
23 Wireless World, 1 September 1926, p. 291
24 Wireless World, 15 September 1926, p. 390
25 Wireless World, 27 October 1926, p. Adv. 4
26 Wireless World, 17 November 1926, p. Adv. 5
27 Wireless World, 1 December 1926, back cover
28 Wireless World, 26 January 1927, p. 111
29 Wireless World, 13 April 1927, p. 469
30 Wireless World, 23 February 1927, pp. 233ff
31 Wireless World, 25 May 1927, p. Adv. 3
32 Wireless World, 26 January 1927, pp. 103ff
33 Wireless World, 8 June 1927, p. 711
34 Wireless World, 19 October 1927, back cover
35 Wireless World, 28 September 1927, p. 414
36 Wireless World, 5 October 1927, p. Adv. 12
37 Wireless World, 26 October 1927, p. Adv. 16
38 Wireless World, 26 October 1927, p. Adv. 10
39 Wireless World, 21 December 1927, p. Adv. 1
40 Wireless World, 18 July 1928, editorial
41 Wireless World, 22 February 1928, p. 200
42 Wireless World, 19 September 1928, p. Adv. 26
43 Wireless World, 26 September 1928, p. 411
44 Wireless World, 31 October 1928, p. Adv. 22
45 Wireless World, 19 September 1928, p. Adv. 24
46 Wireless World, 31 October 1928, p. Adv. 1
47 Wireless World, 31 October 1928, p. Adv. 21
48 Wireless World, 7 November 1928, p. Adv. 12
49 Wireless World, 14 November 1928, p. Adv. 32
50 Wireless World, 21 November 1928, p. Adv. 34
51 Wireless World, 3 October 1928, pp. 461ff
52 Wireless World, 23 January 1929, p. 92
53 Wireless World, 1 October 1930, p. Adv. 34
54 Wireless World, 6 August 1930, p. 216
55 Wireless World, 13 August 1930, p. 149
56 Wireless World, 15 October 1930, p. 454

57 Wireless World, 11 September 1929, p. 236
58 B.B.C. Year Book, 1931, p. 354
59 Modern Wireless, November 1930, p. 463
60 Wireless World, 17 September 1930, p. 281
61 Wireless World, 11 February 1931, p. 160
62 Wireless World, 25 February 1931, p. 220
63 Wireless World, 11 March 1931, p. 270
64 The Wireless Constructor, October 1931, p. 279
65 Wireless World, 30 September 1931, p. 388
66 The Wireless Constructor, October 1931, p. 282
67 Wireless World, 19 August 1932, p. 160
68 The Wireless Constructor, January 1933, p. 181
69 Wireless World, 18 August 1933, p. 127
70 B.B.C. Year Book, 1934, p. 427
71 Wireless World, 17 August 1934, p. 121
72 Wireless World, 24 August 1934, p. 171
73 Wireless World, 16 February 1934, editorial
74 Wireless World, 16 February 1934, pp. 117ff
75 Practical and Amateur Wireless, 29 August 1936, p. 603
76 Practical and Amateur Wireless, 29 August 1936, p. 617
77 Practical and Amateur Wireless, 5 September 1936, p. 682
78 Practical and Amateur Wireless, 28 November 1936, p. 331
79 Practical and Amateur Wireless, 26 September 1936, p. 34
80 The Wireless Constructor, February 1932, p. 219
81 The British Vintage Wireless Society Bulletin, Vol. IV, No. 2, p. 24

Bibliography

W. J. Baker: *A History of the Marconi Company*, Methuen & Co. Ltd. (1970)

Asa Briggs: *The History of Broadcasting in the United Kingdom*, Oxford University Press, 4 vols., Vol. I The Birth of Broadcasting (1961), Vol. II The Golden Age of Wireless (1965), Vol. III The War of Words (1970), Vol. IV Sound and Vision (1979)

Asa Briggs: *The BBC: The First Fifty Years*, Oxford University Press (1985)

Susan Briggs: *Those Radio Times*, Weidenfeld and Nicolson (1981)

Gordon Bussey: See entries *Dictionary of Business Biography*, Butterworths, 5 vols., Vol. I E. K. Cole (1985), Vol. IV H. J. Pye (1985), Vol. IV W. G. Pye (1985)

Gordon Bussey: *The Story of Pye Wireless*, Pye Limited (1979)

Gordon Bussey, (Ed.): Supplement to Science Museum Exhibition, *The Great Optical Illusion*, Philips Industries (1980)

Gordon Bussey, (Ed.): Supplement to V & A Exhibition, *'The Wireless Show!'*, Philips Industries (1977)

Gordon Bussey: *Vintage Crystal Sets 1922–1927*, IPC Electrical-Electronic Press Ltd (1976)

Anthony Constable: *Early Wireless*, Midas Books (1980)

Keith Geddes: *Broadcasting in Britain*, Science Museum (1972)

Keith Geddes & Gordon Bussey: *The History of Roberts Radio*, Roberts Radio Co. Ltd. (1988)

Keith Geddes in collaboration with Gordon Bussey: *The Setmakers*, British Radio & Electronic Equipment Manufacturers' Association (1991)

Keith Geddes & Gordon Bussey: *Television the first fifty years*, National Museum Photography, Film & Television (1986)

Jonathan Hill: *The Cat's Whisker*, 50 Years of Wireless Design, Oresko Books Ltd (1978)

Jonathan Hill: *Radio! Radio!*, Sunrise Press (1986)

Joan Long: *A First Class Job!*, Frank Murphy – radio pioneer, The author (1985)

R. F. Pocock: *The Early British Radio Industry*, Manchester University Press (1988)

Tim Wander: *2MT Writtle*, The Birth of British Broadcasting, Capella Publications (1988)

Norman Wymer: *Guglielmo Marconi*, GEC-Marconi Electronics Limited (c.1980)

Index